THE HISTORICAL CATALOGUE SERIES

"Whiting Manufacturing Company -- 1896"
Catalogue Reprint

First Printing

Edited by
Benjamin A. Randolph

Published by
The Eden Sterling Company
7672 Montgomery Road
Cincinnati, Ohio 45236
800-385-3336

Other Publications in the Historical Catalogue Series
"Gorham Strasbourg - 1910"
"Gorham Buttercup - 1910"
"Gorham Sterling Flatware - 1888"
"The Colonial Book of the Towle Manufacturing Company - 1898"
"The Colonial Book of the Towle Manufacturing Company - 1910"

Note: In order to maintain the authenticity of this catalogue, the images have not been retouched and some may appear faint.

ISBN 0-9656139-2-5

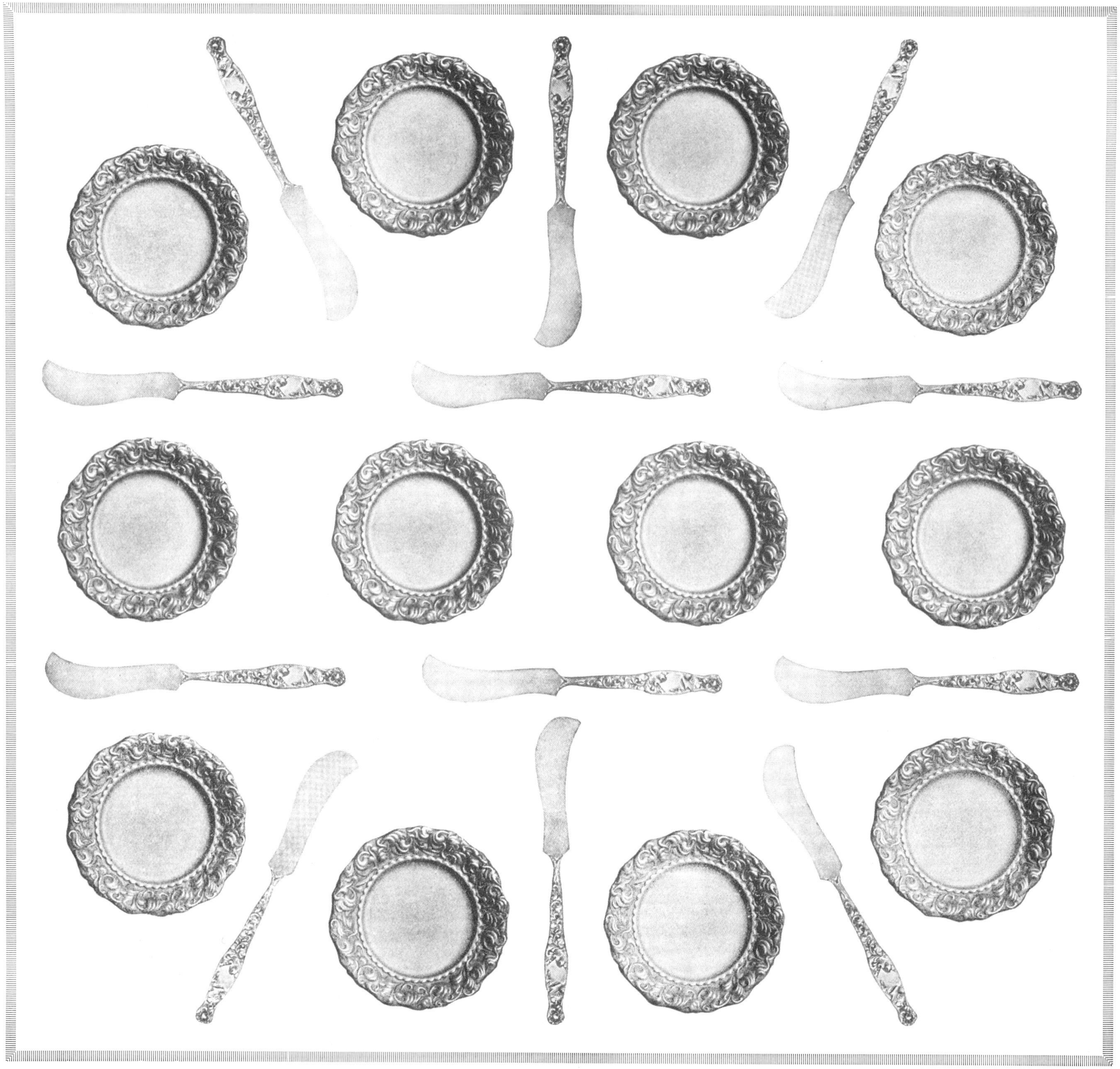

62 A 40

(Copyrighted.)

WHITING M'F'G CO., SILVERSMITHS,

BROADWAY & 18TH ST., NEW YORK.

(Half Actual Size.)

90 A 9

90 A 17

WHITING M'F'G CO., BROADWAY & 18TH ST.,

(COPYRIGHTED.) SILVERSMITHS, TRADE MARK STERLING NEW YORK. (HALF ACTUAL SIZE.)

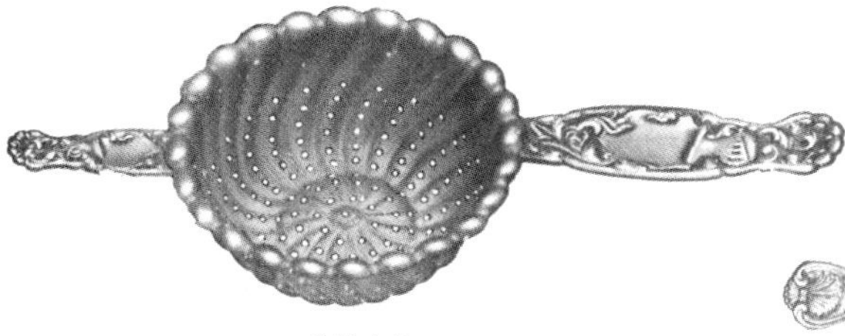

245 A 1

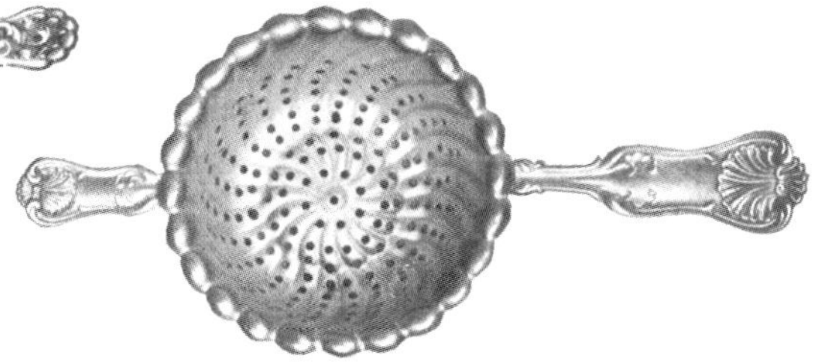

245 A 5

245 A 8

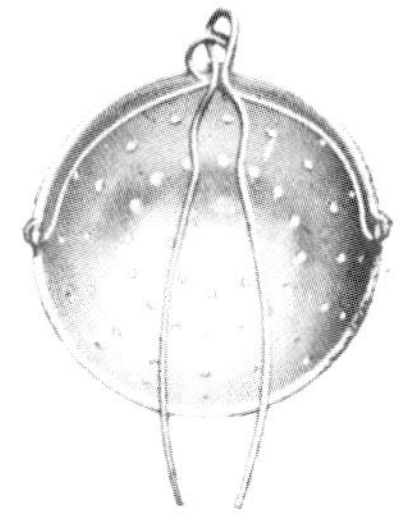

245 A 9

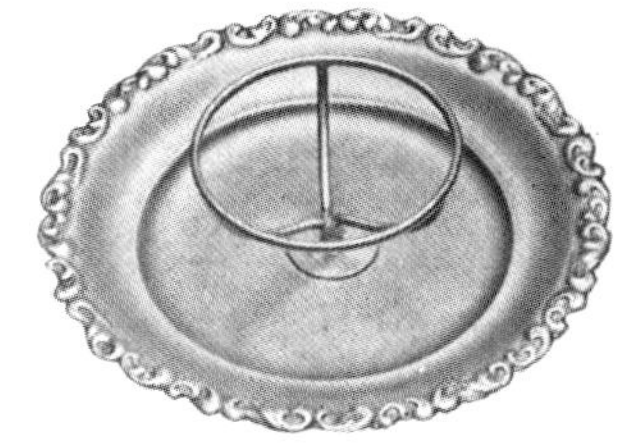

245 A 24

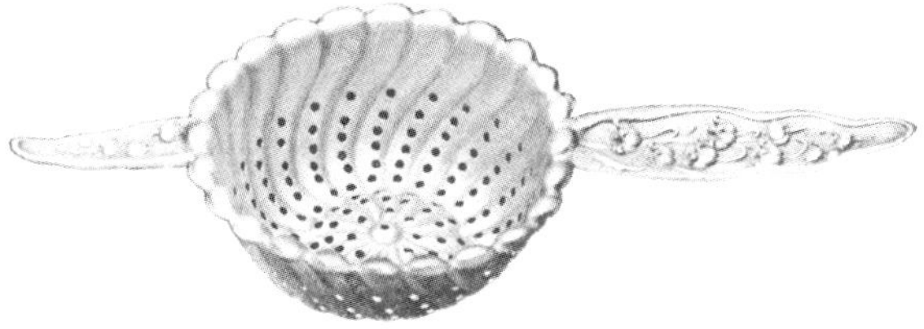

245 A 2

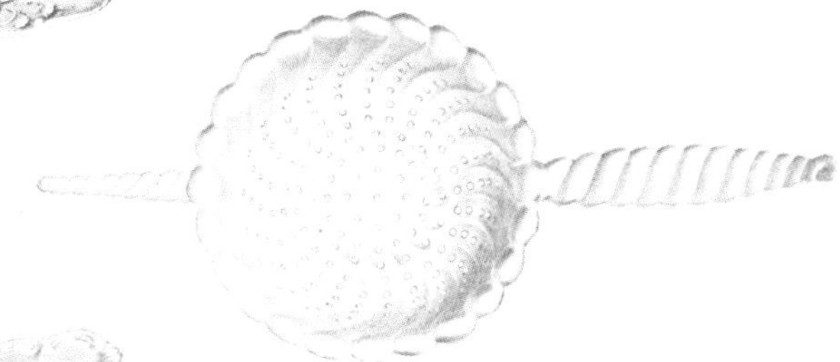

245 A 6

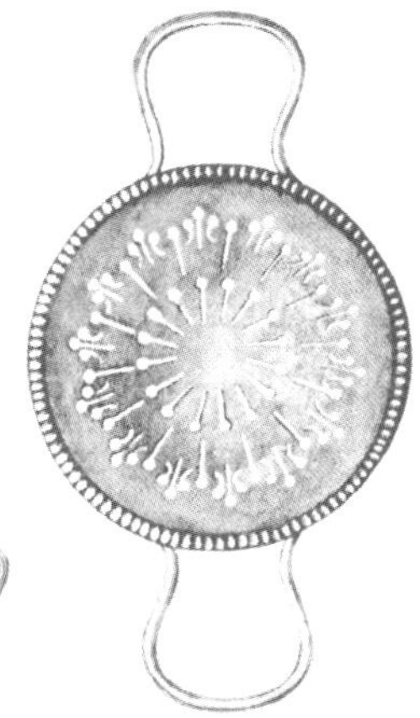

245 A 10

245 A 11

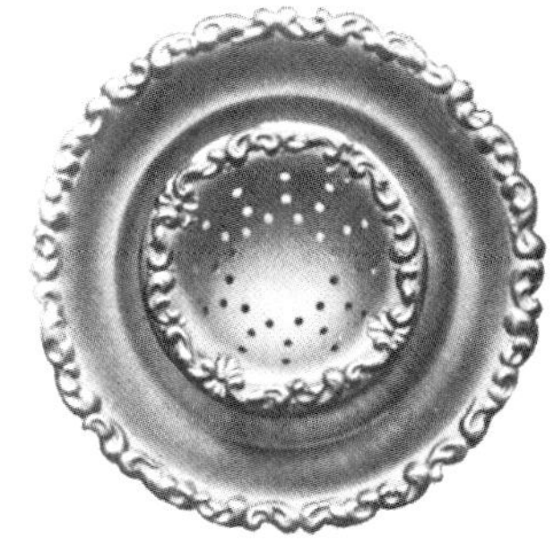

245 A 25

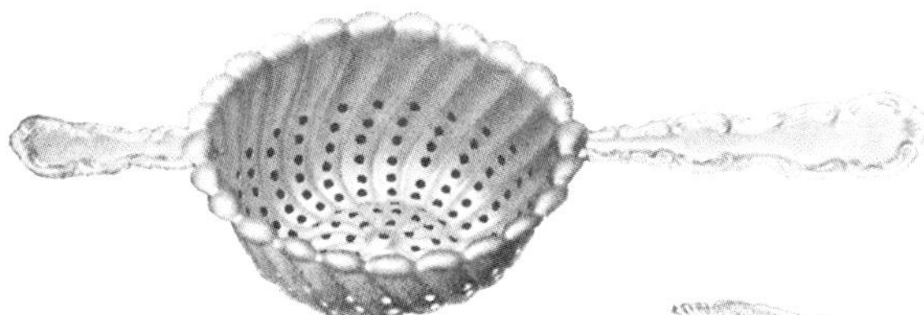

245 A 3

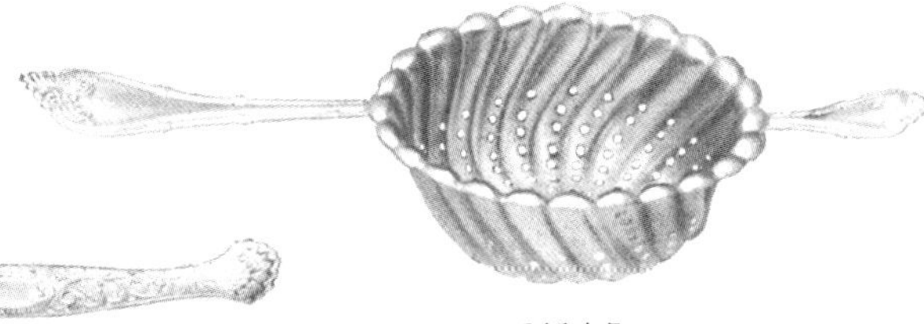

245 A 7

245 A 12

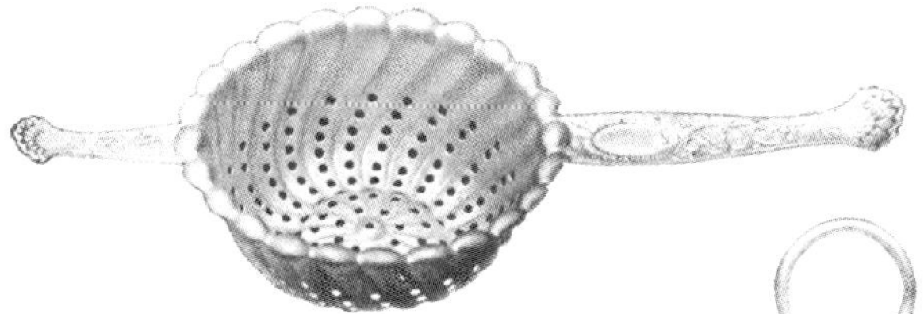

245 A 4

245 A 13

245 A 26

245 A 14

245 A 15

245 A 16

245 A 17

245 A 18

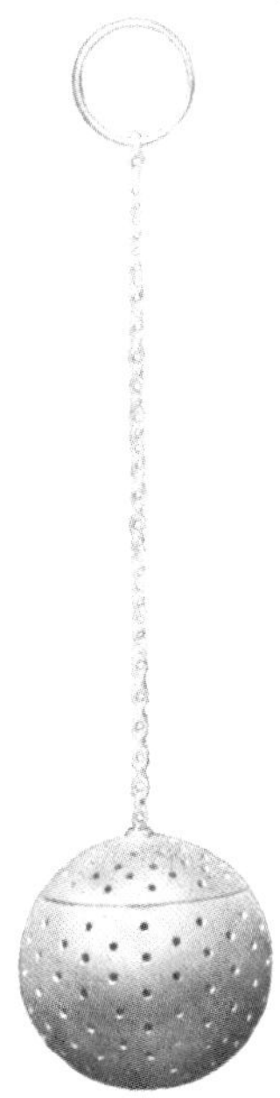

245 A 19

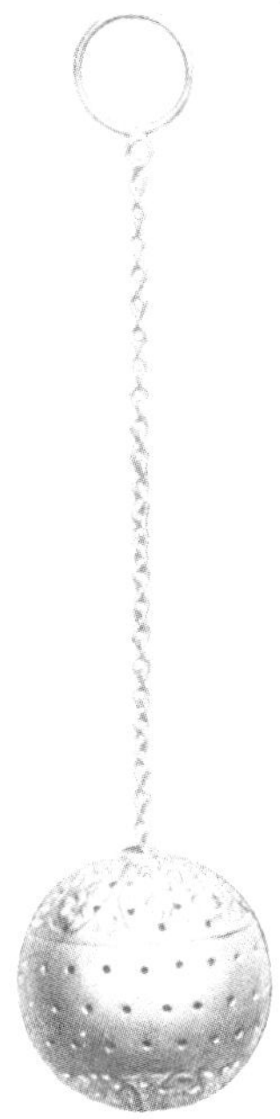

245 A 20

245 A 21

245 A 22

245 A 23

WHITING M'F'G CO., SILVERSMITHS, BROADWAY & 18TH ST., NEW YORK.

(HALF ACTUAL SIZE.)

248 A 43

248 A 44

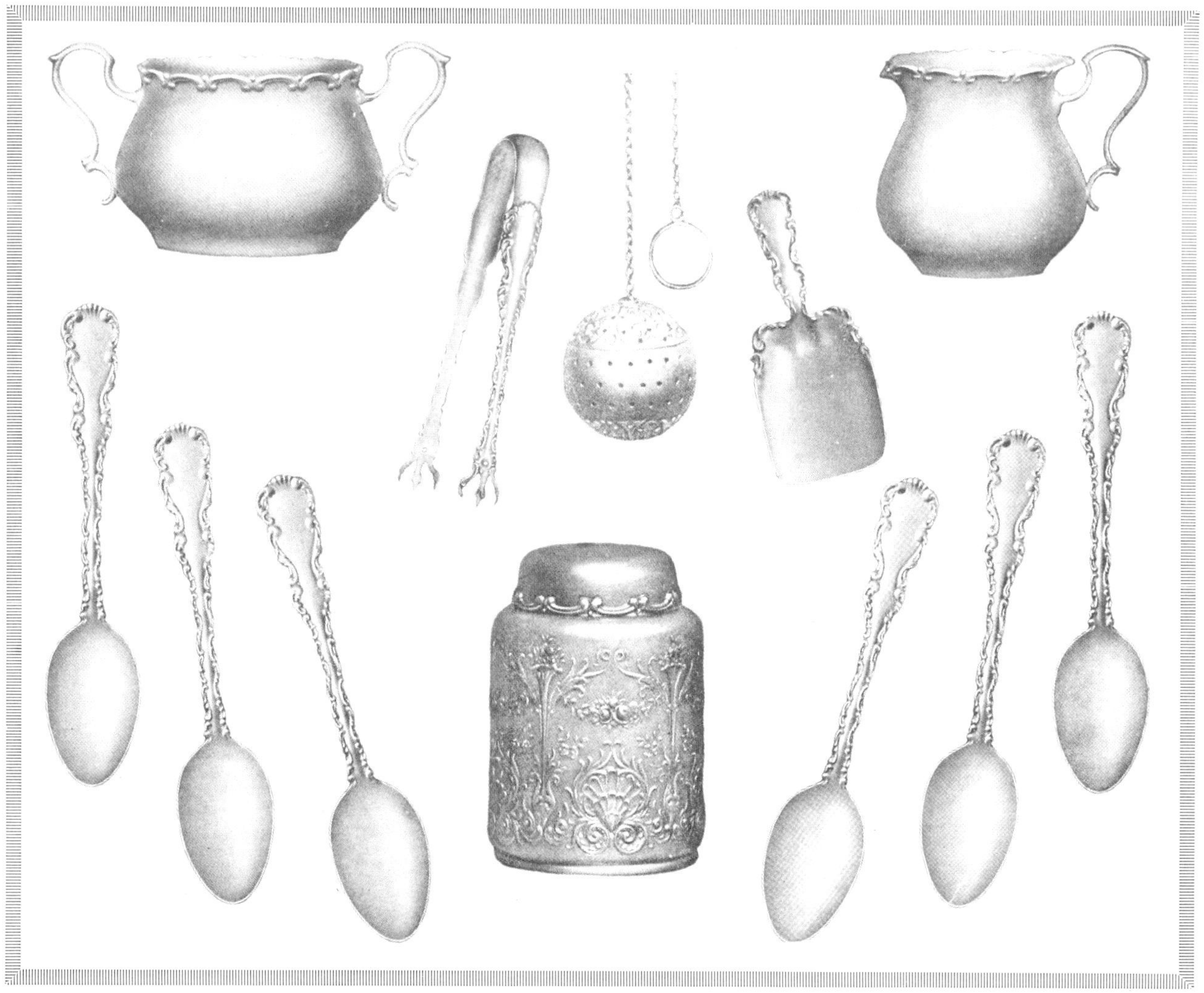

248 A 45

248 A 79

(HALF ACTUAL SIZE.)

248 A 87

WHITING M'F'G CO., SILVERSMITHS, BROADWAY & 18TH ST., NEW YORK.

TRADE MARK STERLING

 (HALF ACTUAL SIZE.)

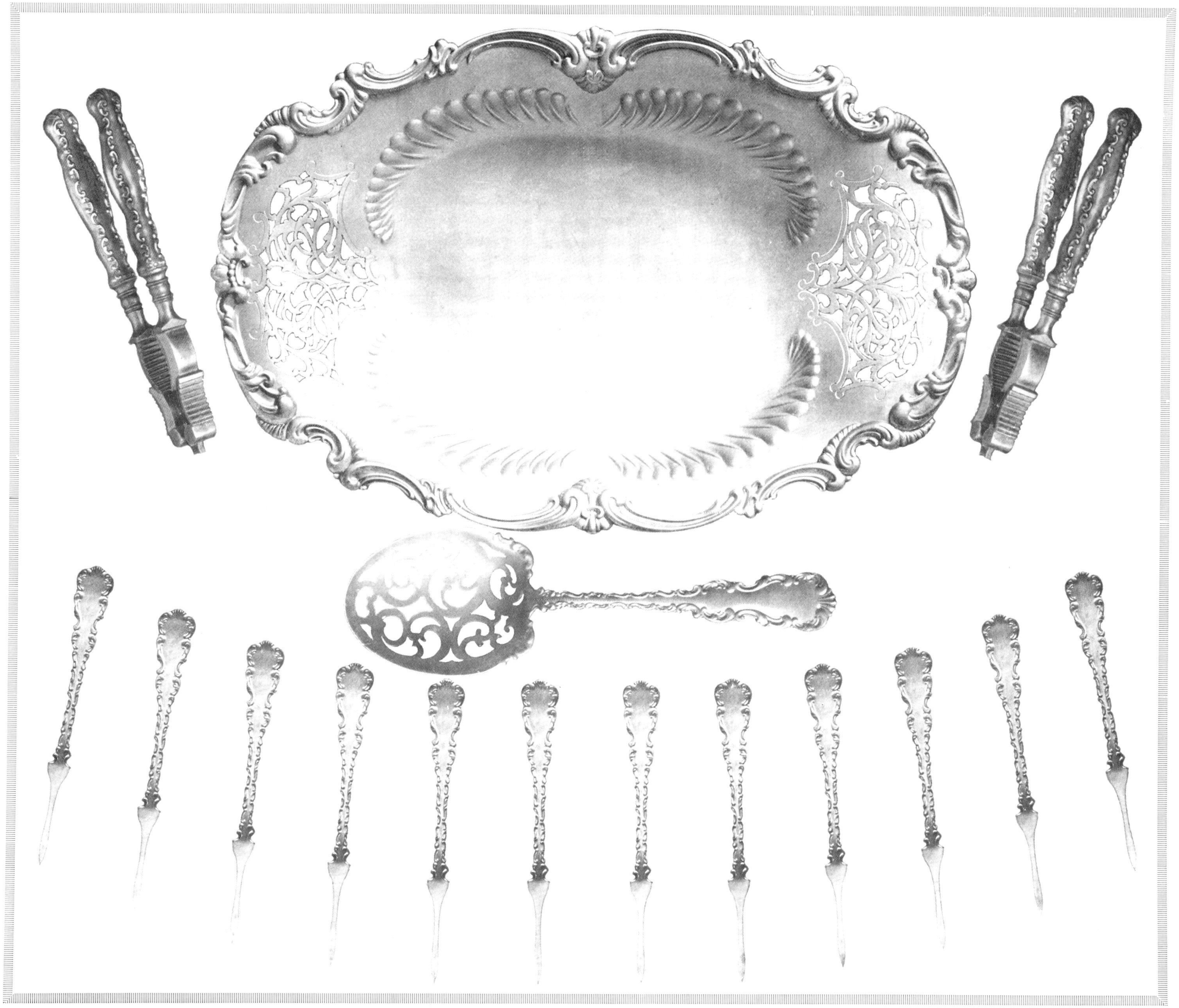

345 A 125

WHITING M'F'G CO., SILVERSMITHS,

BROADWAY & 18TH ST., NEW YORK.

TRADE MARK STERLING

(HALF ACTUAL SIZE.)

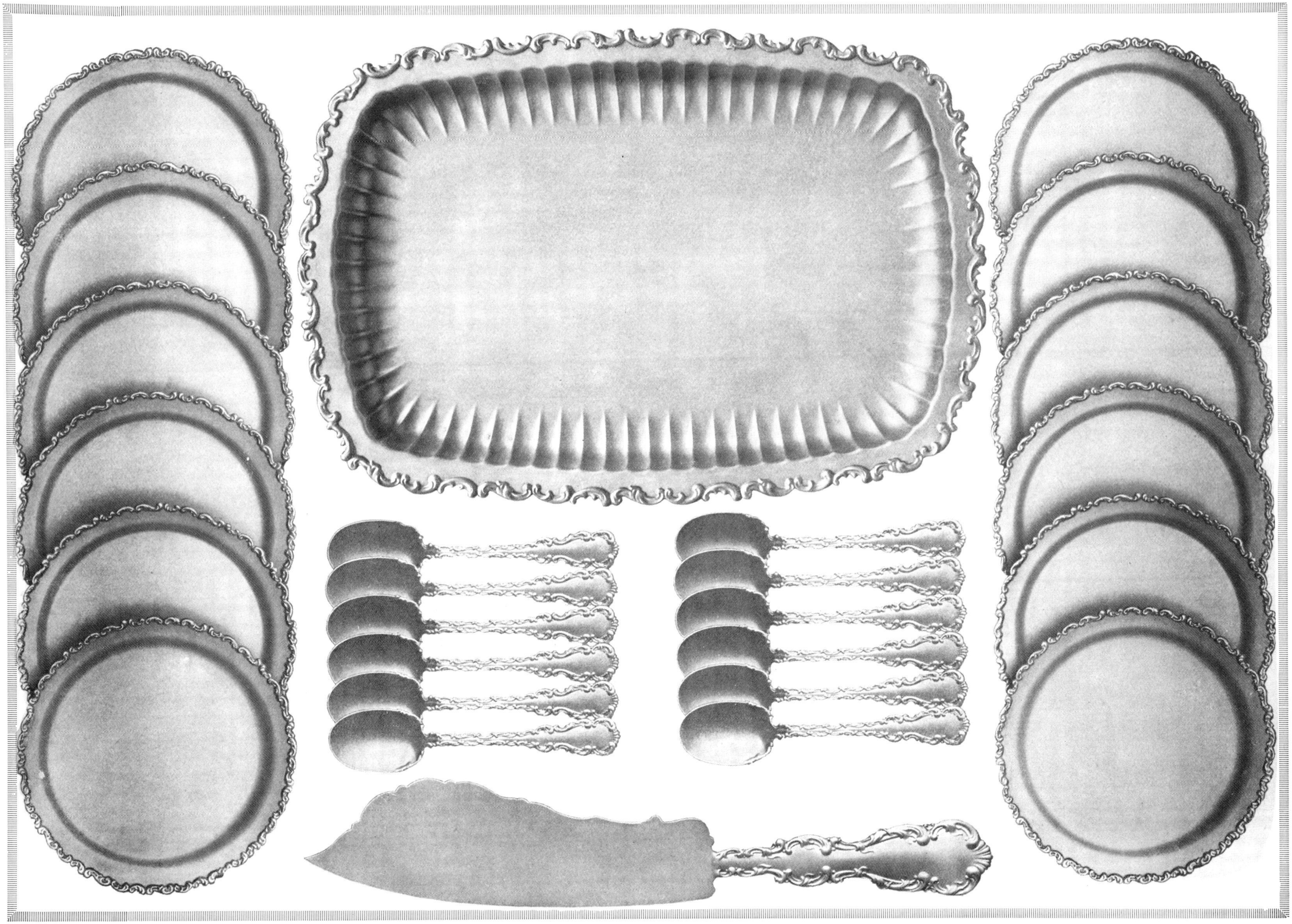

345 A 149

WHITING M'F'G CO., SILVERSMITHS, BROADWAY & 18TH ST., NEW YORK.

TRADE MARK W STERLING

(HALF ACTUAL SIZE.)

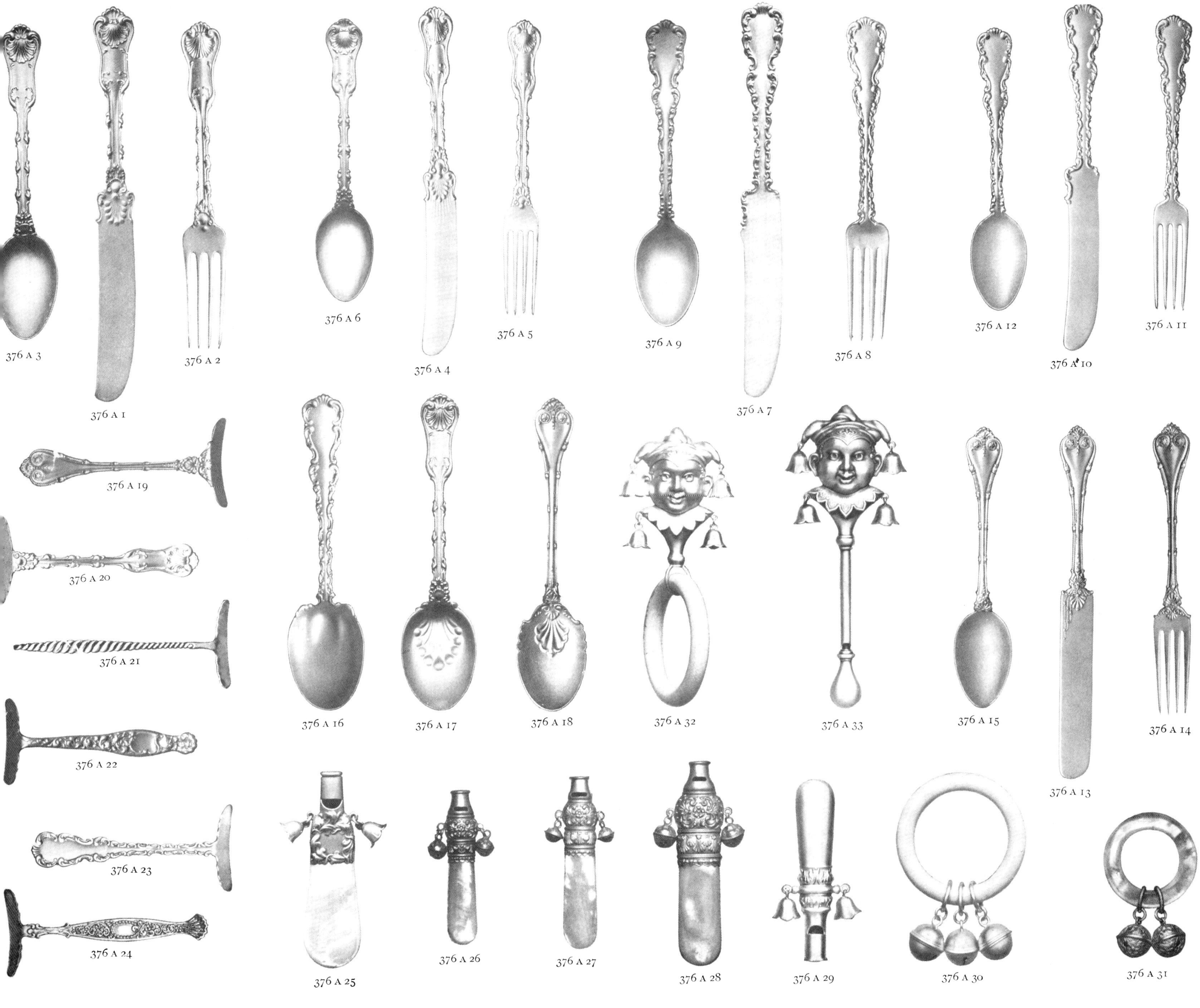

WHITING M'F'G CO.,
SILVERSMITHS,

TRADE W MARK
STERLING

BROADWAY & 18TH ST.,
NEW YORK.

(HALF ACTUAL SIZE.)

380 A 2

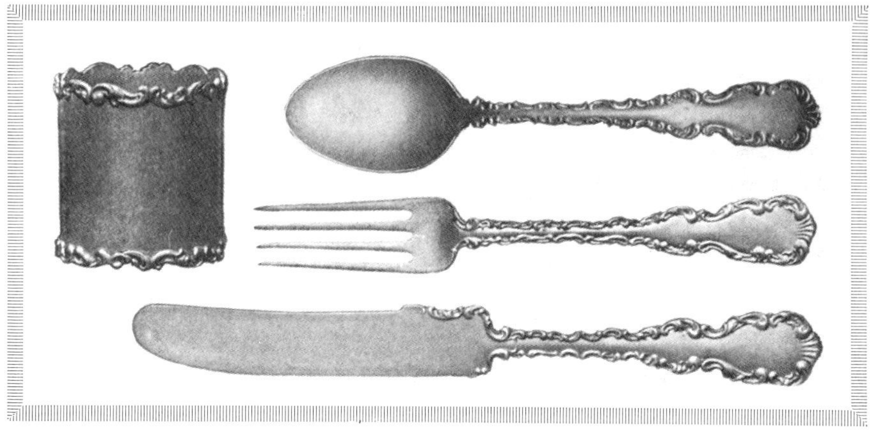

380 A 1

380 A 3

380 A 4

380 A 31

WHITING M'F'G CO., SILVERSMITHS, BROADWAY & 18TH ST., NEW YORK.

TRADE MARK STERLING

(HALF ACTUAL SIZE.)

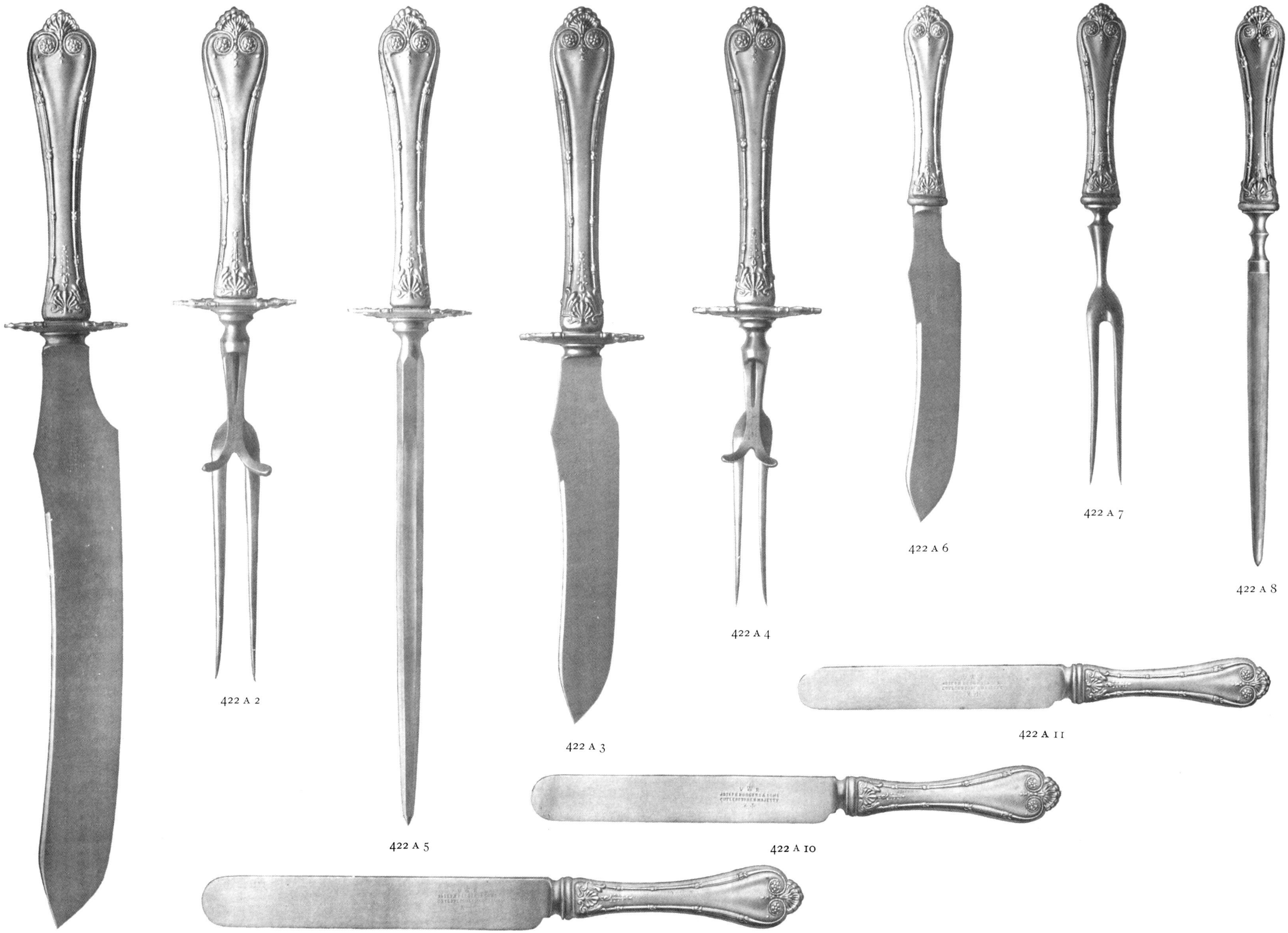

422 A 1 422 A 2 422 A 3 422 A 4 422 A 5 422 A 6 422 A 7 422 A 8 422 A 9 422 A 10 422 A 11

WHITING M'F'G CO., BROADWAY & 18TH ST.,

(COPYRIGHTED.) SILVERSMITHS, NEW YORK. (HALF ACTUAL SIZE.)

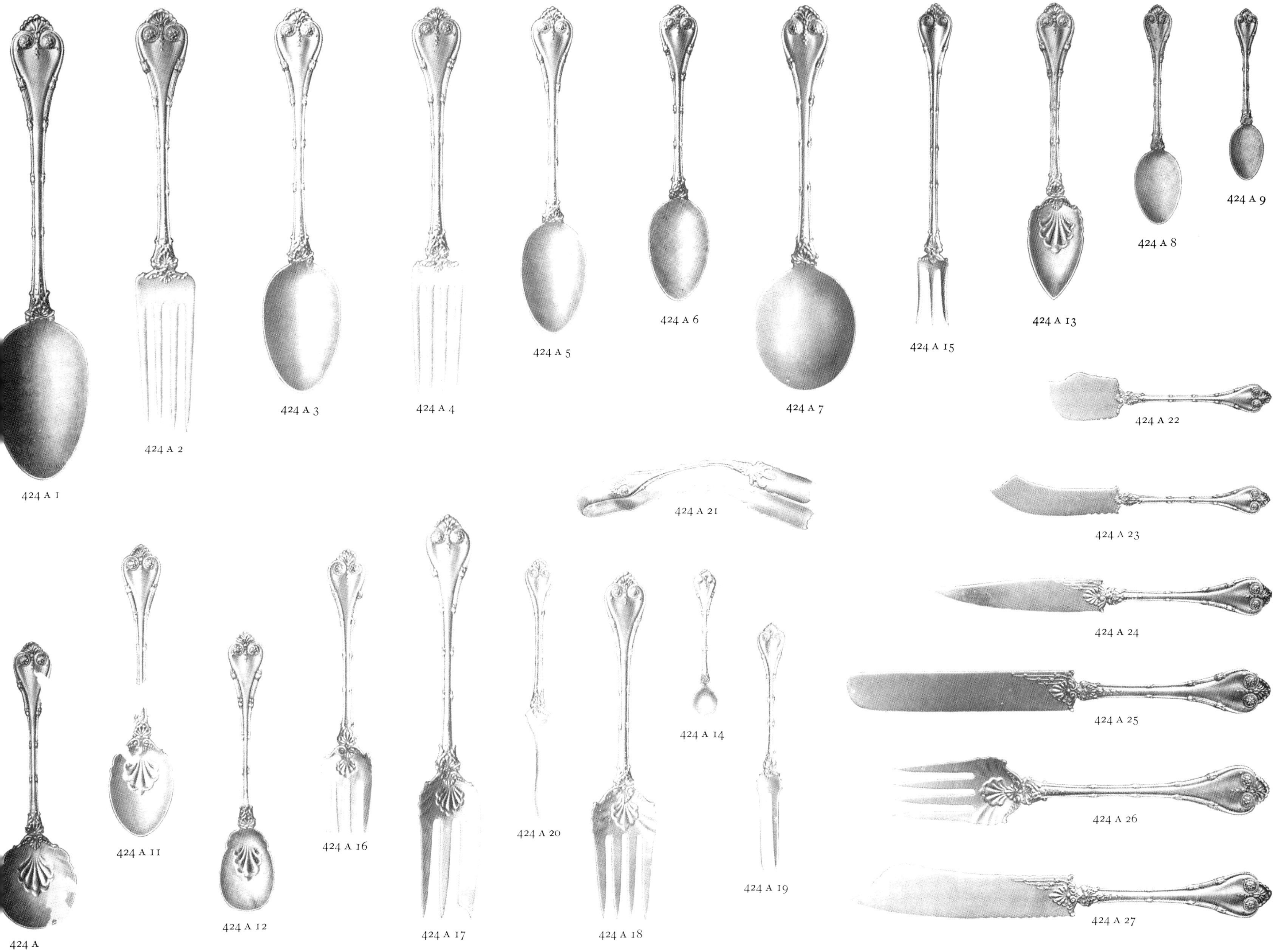

WHITING M'F'G CO., SILVERSMITHS,

TRADE MARK STERLING

BROADWAY & 18TH ST., NEW YORK.

(HALF ACTUAL SIZE.)

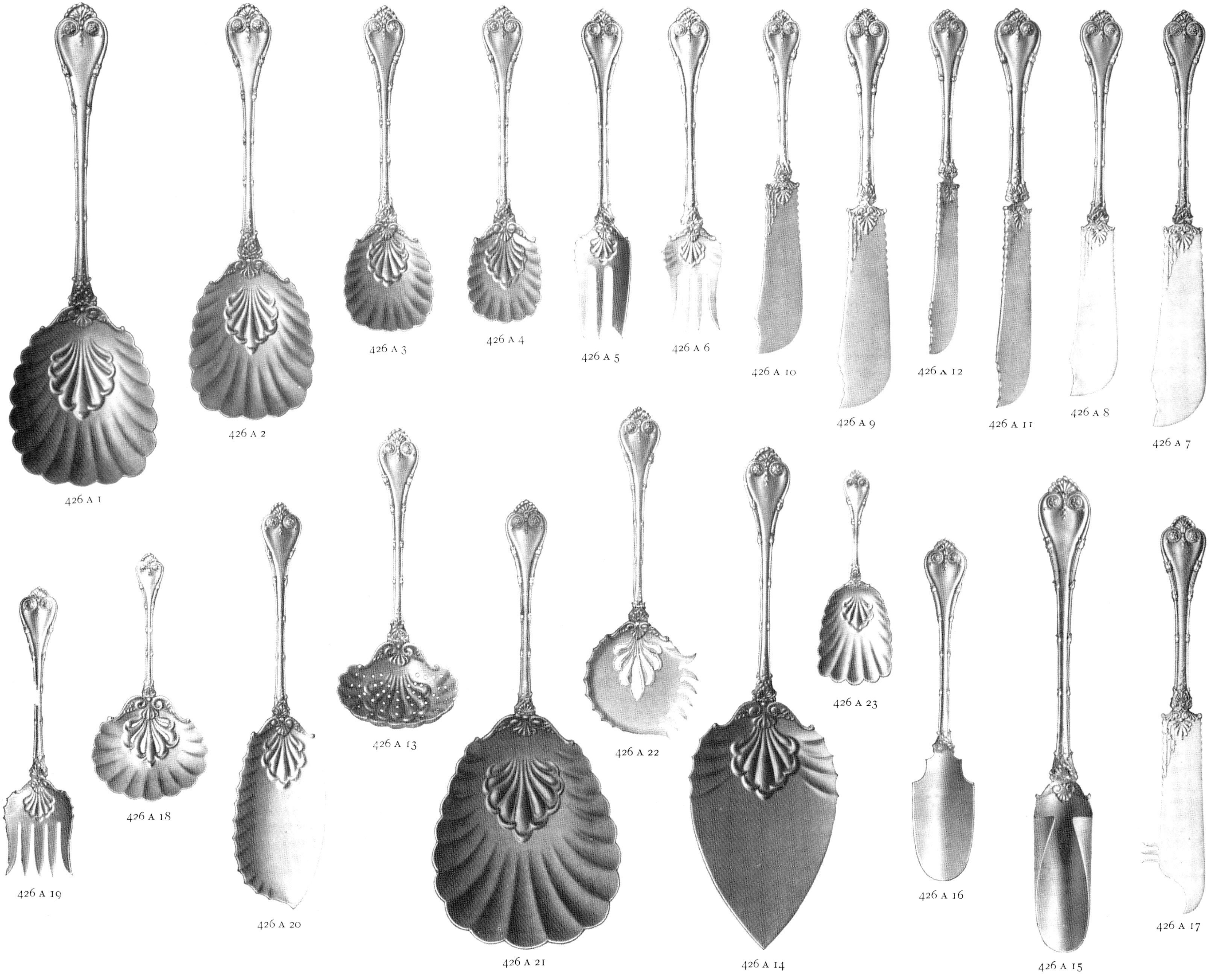

426 A 1 426 A 2 426 A 3 426 A 4 426 A 5 426 A 6 426 A 10 426 A 9 426 A 12 426 A 11 426 A 8 426 A 7

426 A 19 426 A 18 426 A 20 426 A 13 426 A 21 426 A 22 426 A 14 426 A 23 426 A 16 426 A 15 426 A 17

WHITING M'F'G CO., BROADWAY & 18TH ST.,

SILVERSMITHS, NEW YORK.

TRADE MARK STERLING

(HALF ACTUAL SIZE.)

428 A 1
428 A 2
428 A 3
428 A 4
428 A 5
428 A 6
428 A 7
428 A 8
428 A 9
428 A 10
428 A 11
428 A 12
428 A 13

WHITING M'F'G CO., BROADWAY & 18TH ST.,
SILVERSMITHS, NEW YORK.

TRADE MARK STERLING

 (HALF ACTUAL SIZE.)

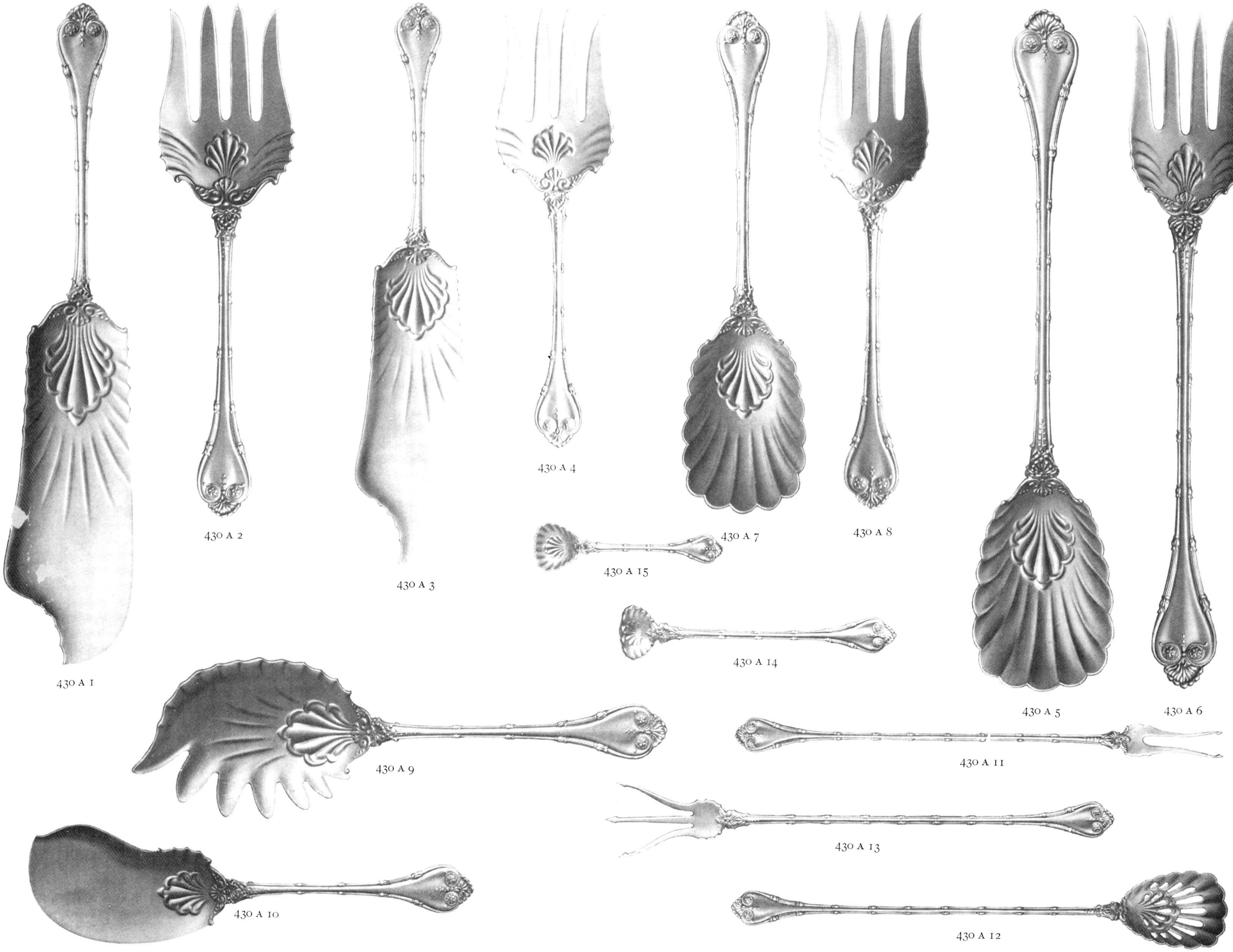

430 A 1, 430 A 2, 430 A 3, 430 A 4, 430 A 5, 430 A 6, 430 A 7, 430 A 8, 430 A 9, 430 A 10, 430 A 11, 430 A 12, 430 A 13, 430 A 14, 430 A 15

(Copyrighted.)

WHITING M'F'G CO.,
SILVERSMITHS,

BROADWAY & 18TH ST.,
NEW YORK.

(Half Actual Size.)

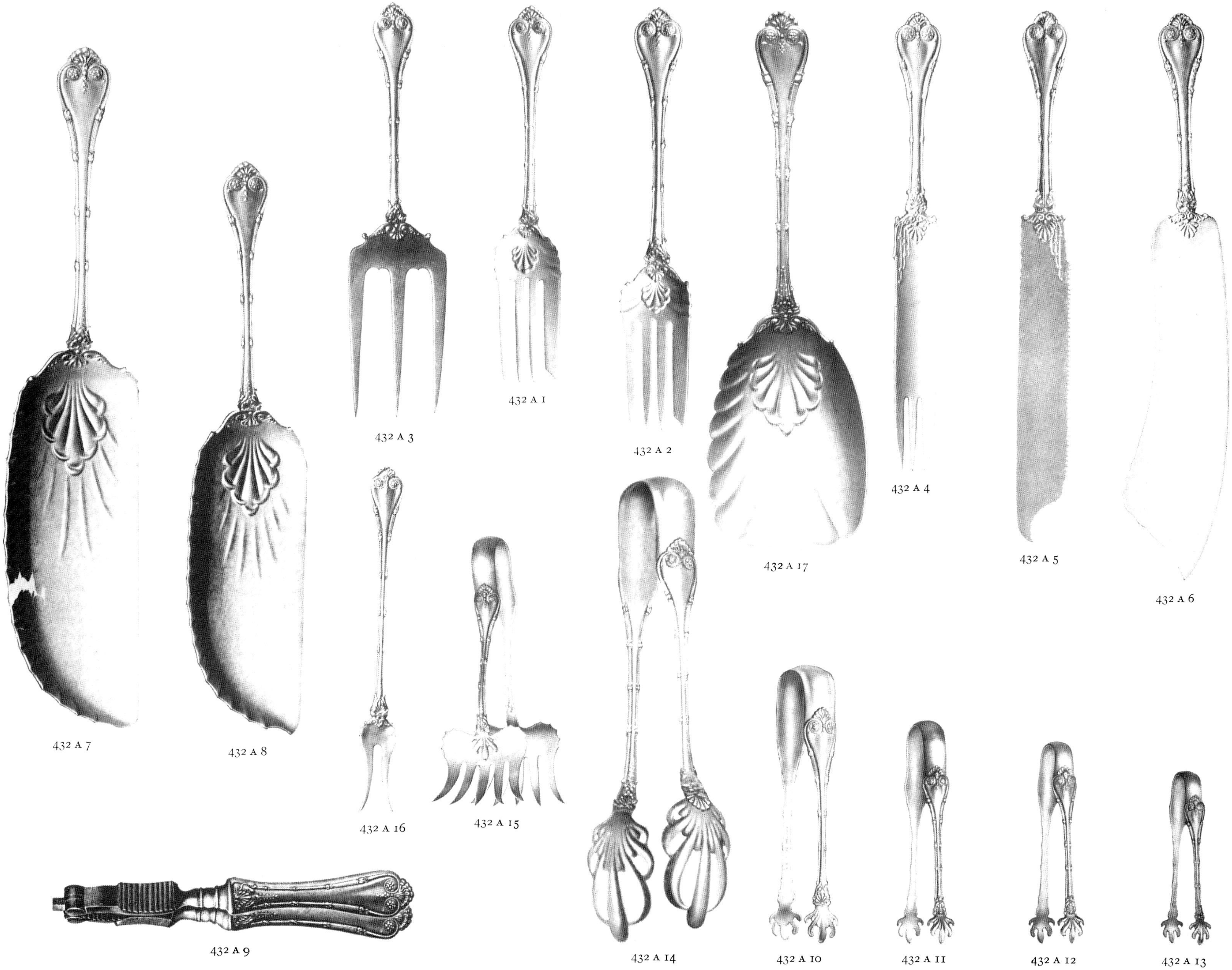

WHITING M'F'G CO., BROADWAY & 18TH ST.,

SILVERSMITHS, NEW YORK.

(HALF ACTUAL SIZE.)

433 A 1

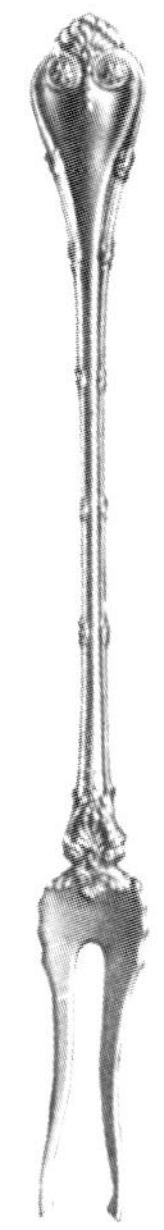

433 A 2

433 A 3

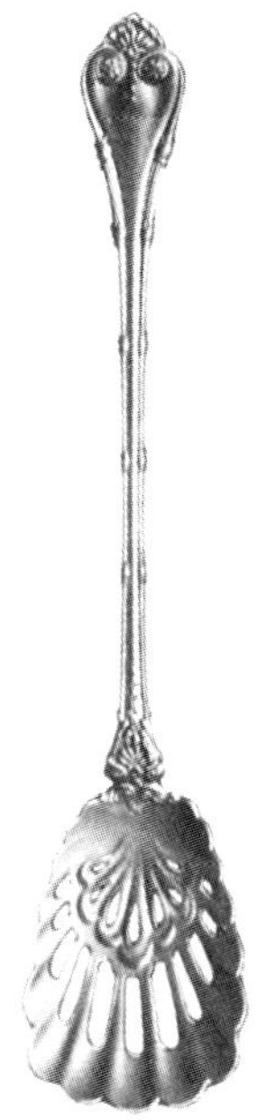

433 A 4

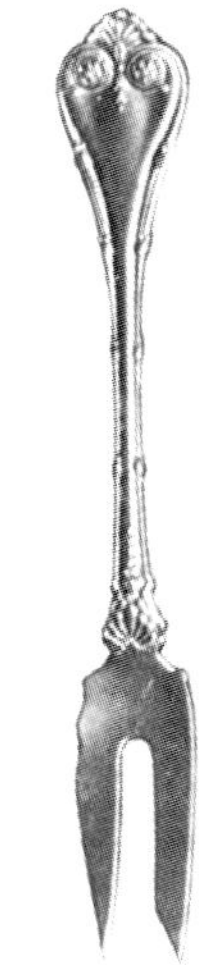

433 A 5

WHITING M'F'G CO., BROADWAY & 18TH ST.,

(COPYRIGHTED.) SILVERSMITHS, NEW YORK. (HALF ACTUAL SIZE.)

440 A 1

440 A 2

440 A 3

440 A 4

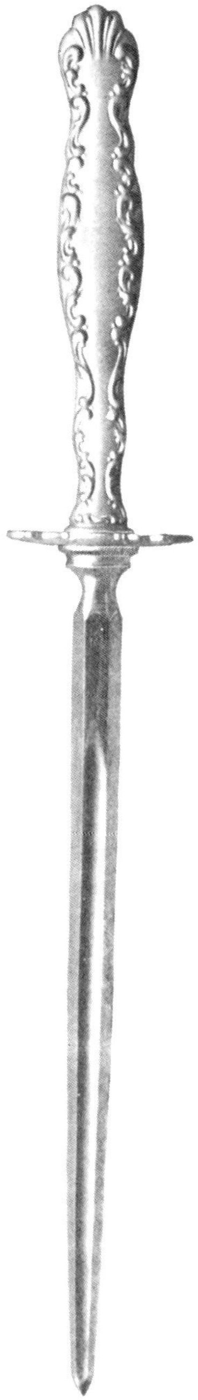

440 A 5

440 A 6

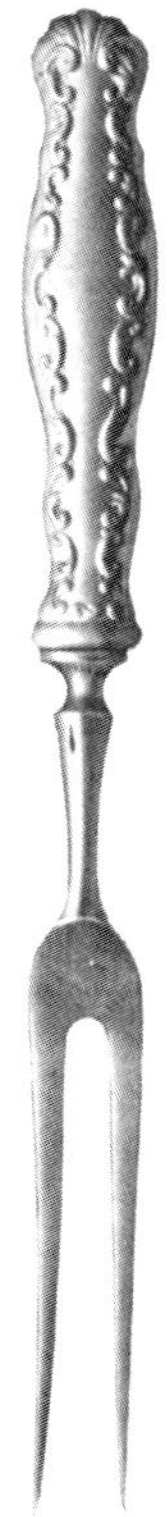

440 A 7

440 A 8

440 A 9

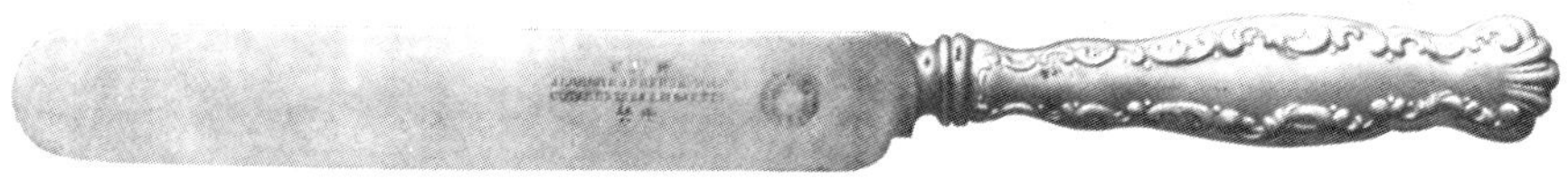

440 A 10

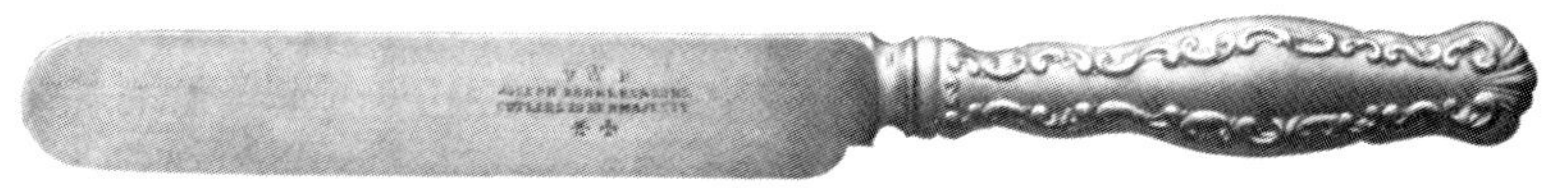

440 A 11

WHITING M'F'G CO., BROADWAY & 18TH ST.,

SILVERSMITHS, NEW YORK.

(HALF ACTUAL SIZE.)

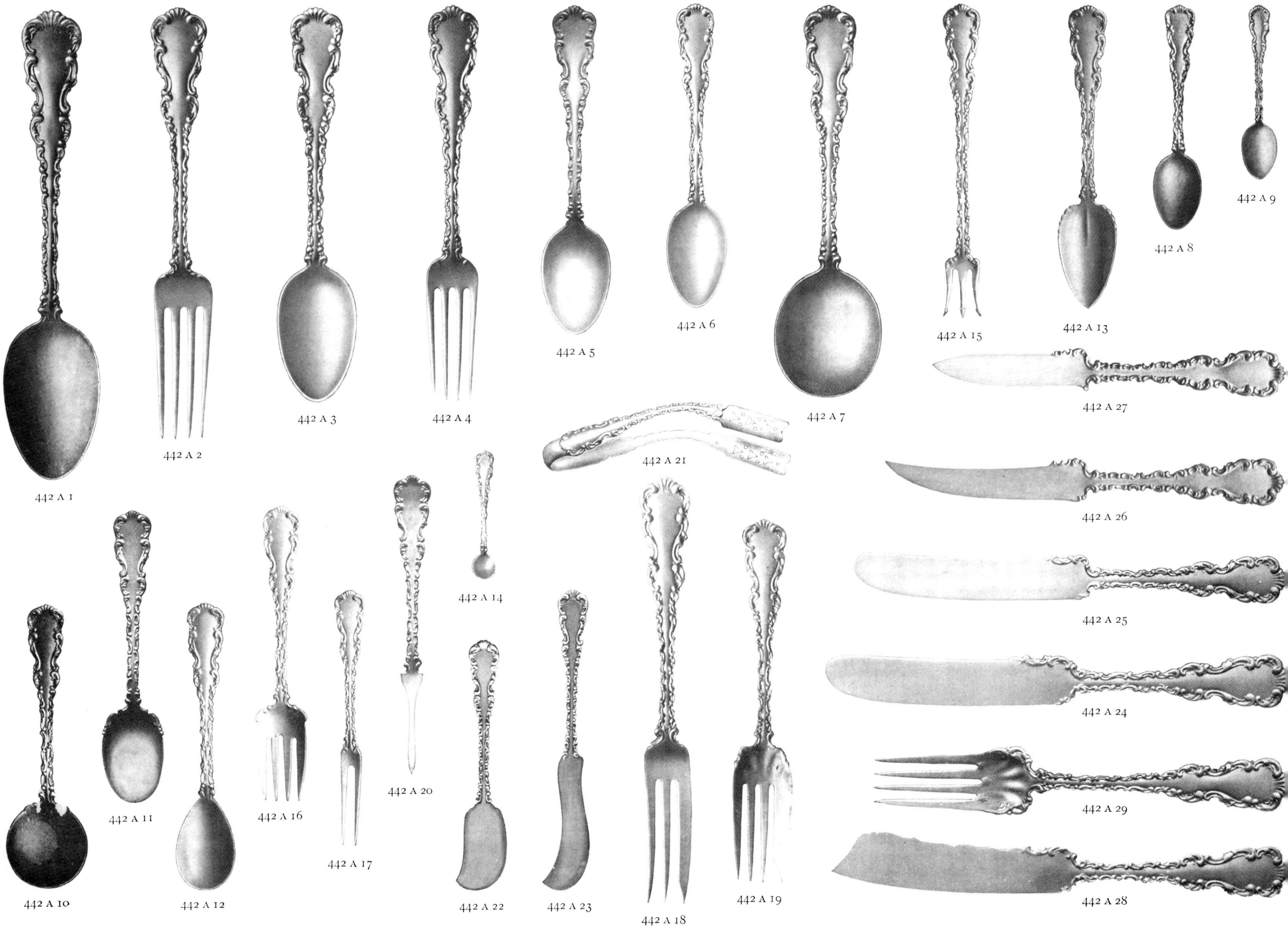

WHITING M'F'G CO., SILVERSMITHS, BROADWAY & 18TH ST. NEW YORK.

TRADE MARK STERLING

(COPYRIGHTED.)

(HALF ACTUAL SIZE.)

444 A 1 · 444 A 2 · 444 A 3 · 444 A 4 · 444 A 5 · 444 A 6 · 444 A 11 · 444 A 10 · 444 A 9 · 444 A 8

444 A 15 · 444 A 21 · 444 A 20

444 A 7 · 444 A 18 · 444 A 17 · 444 A 19 · 444 A 16 · 444 A 14 · 444 A 13 · 444 A 12

WHITING M'F'G CO., BROADWAY & 18TH ST.,

SILVERSMITHS, NEW YORK.

(HALF ACTUAL SIZE.)

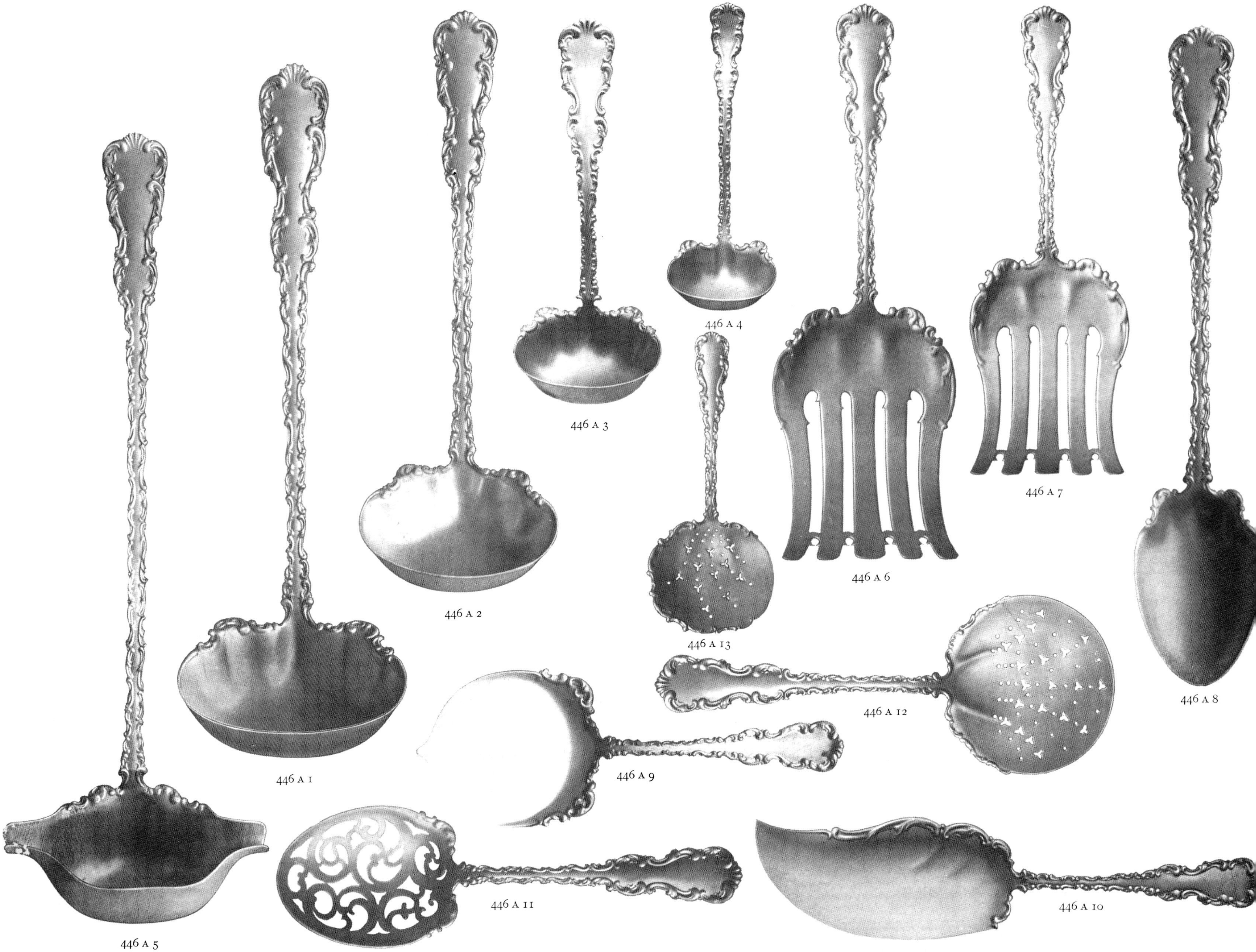

WHITING M'F'G CO., SILVERSMITHS, BROADWAY & 18TH ST., NEW YORK

(COPYRIGHTED.)

(HALF ACTUAL SIZE.)

WHITING M'F'G CO., SILVERSMITHS, BROADWAY & 18TH ST., NEW YORK.

(COPYRIGHTED.)

(HALF ACTUAL SIZE.)

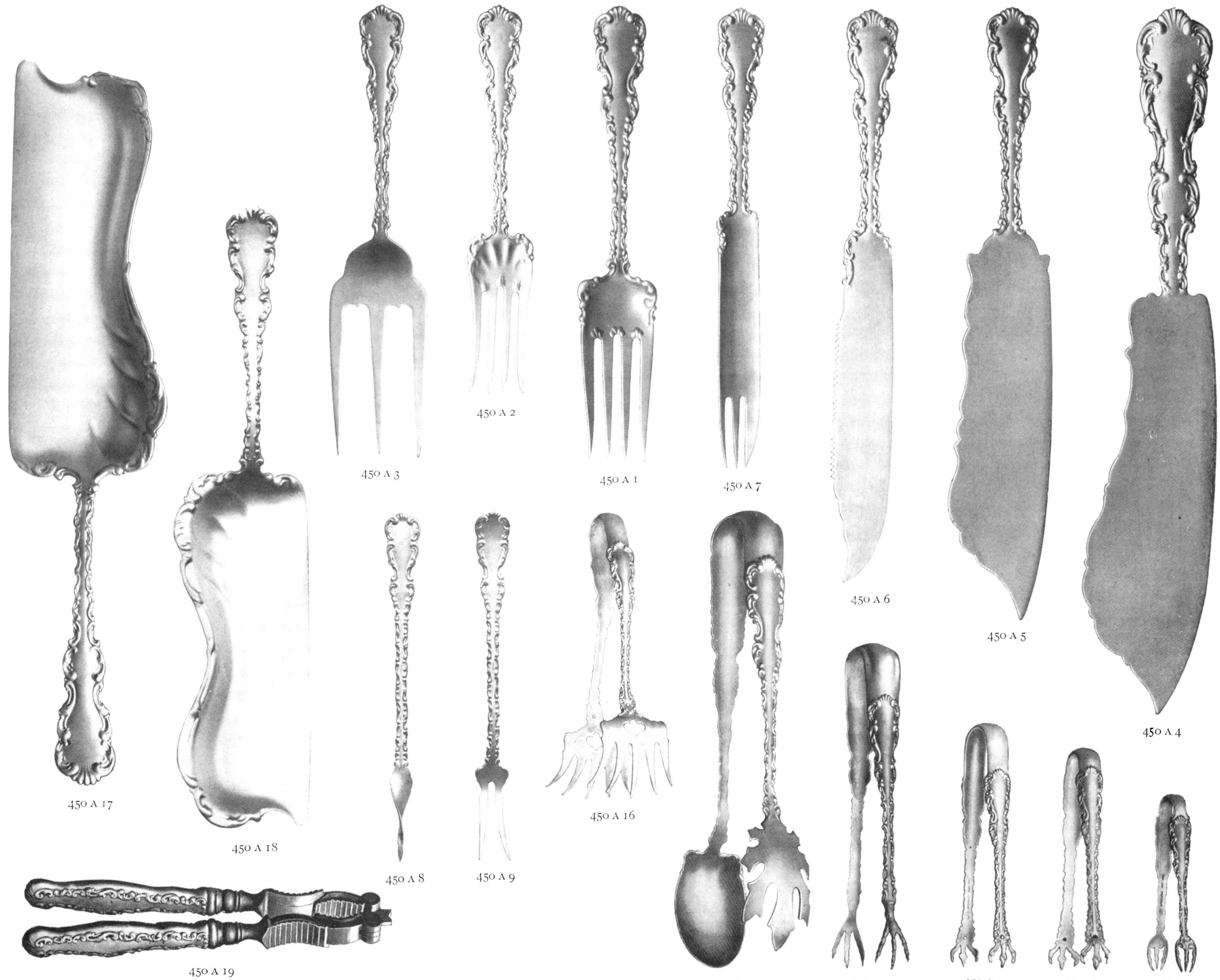

450 A 17, 450 A 18, 450 A 3, 450 A 2, 450 A 1, 450 A 7, 450 A 6, 450 A 5, 450 A 4, 450 A 8, 450 A 9, 450 A 16, 450 A 15, 450 A 10, 450 A 11, 450 A 12, 450 A 13, 450 A 19

WHITING M'F'G CO., SILVERSMITHS, BROADWAY & 18TH ST., NEW YORK.

TRADE MARK STERLING

(HALF ACTUAL SIZE.)

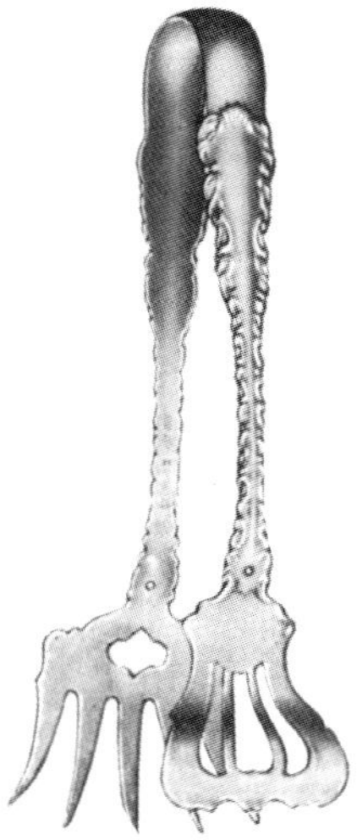

451 A 1

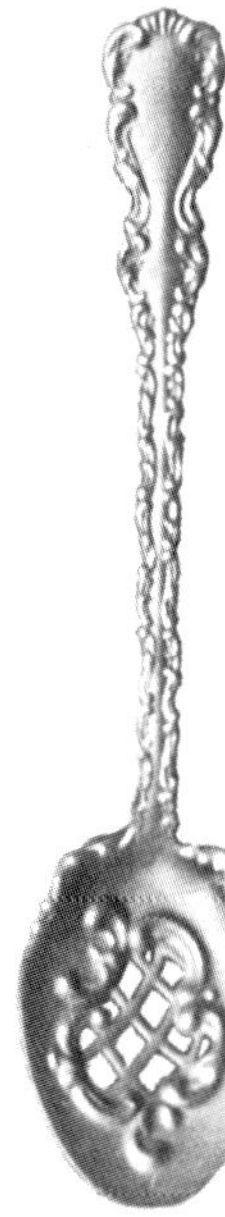

451 A 2

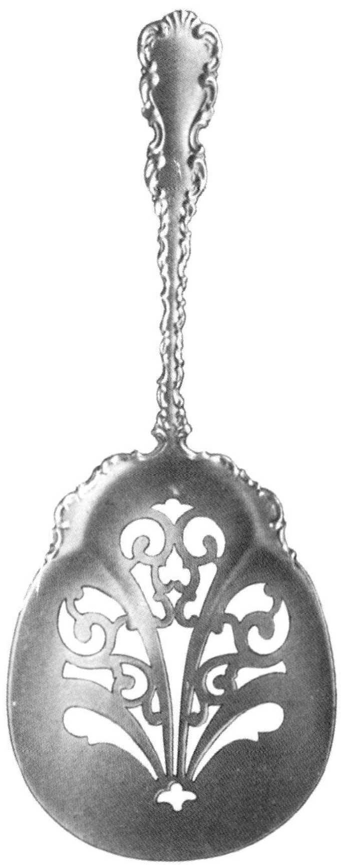

451 A 3

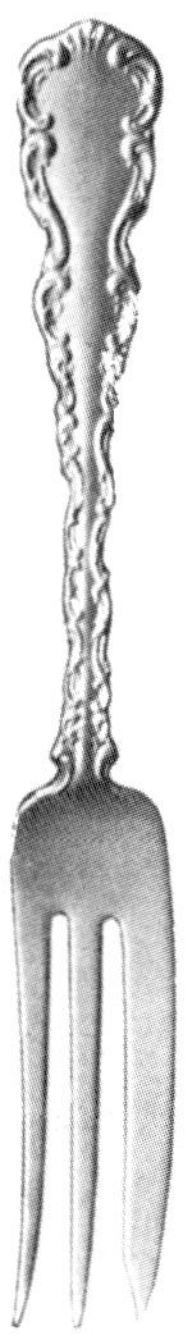

451 A 4

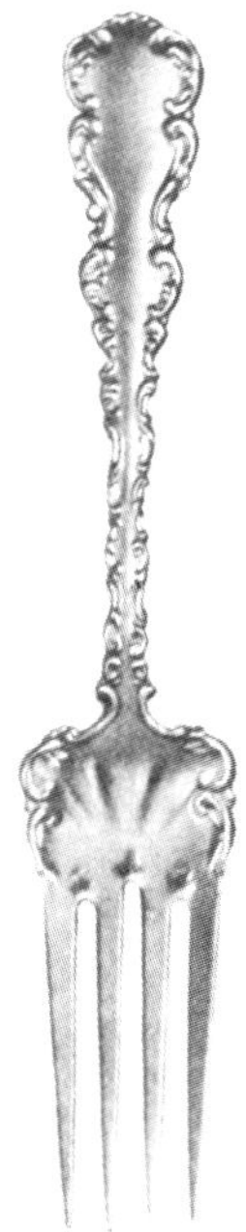

451 A 5

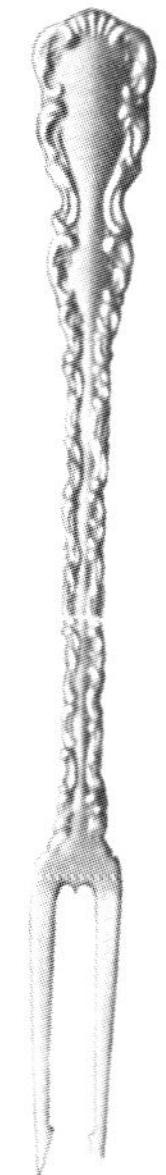

451 A 6

451 A 7

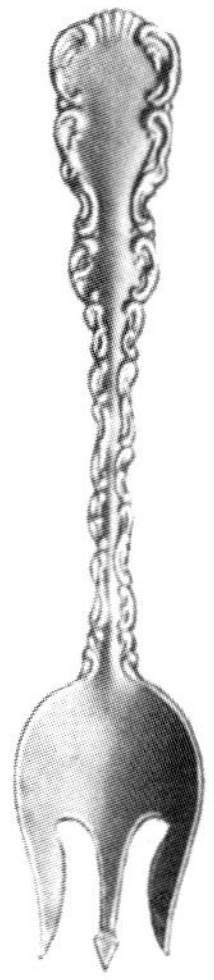

451 A 8

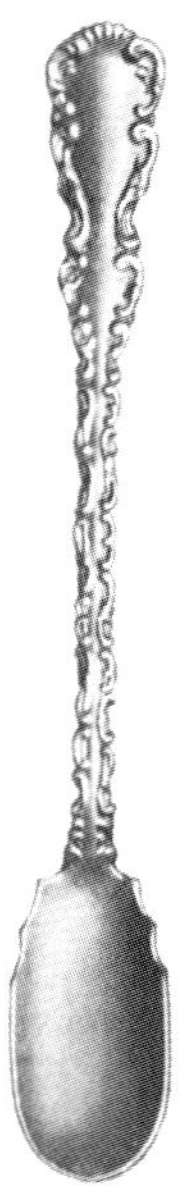

451 A 9

451 A 10

WHITING M'F'G CO., SILVERSMITHS, BROADWAY & 18TH ST. NEW YORK.

TRADE MARK STERLING

(HALF ACTUAL SIZE.)

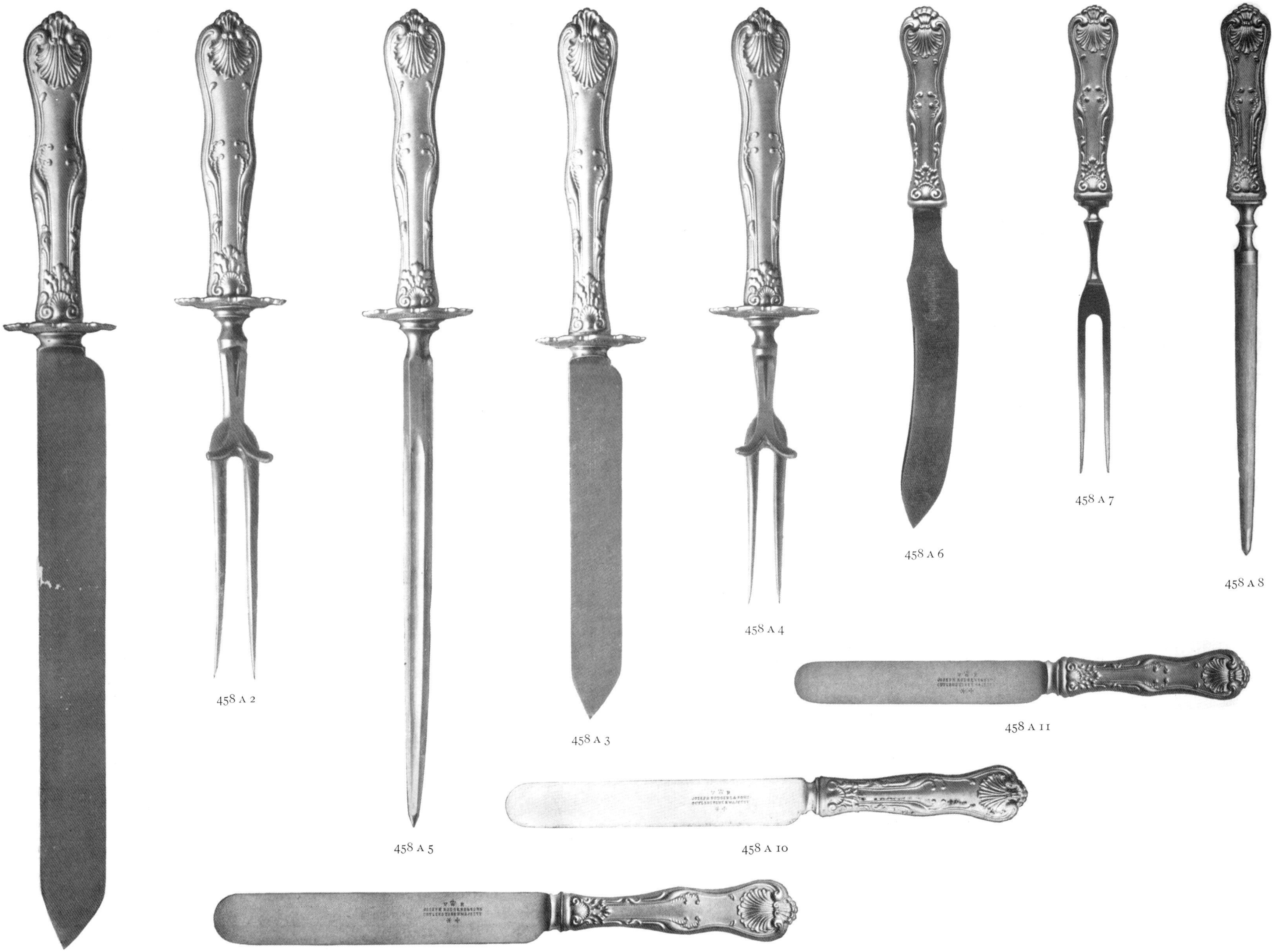

458 A 1, 458 A 2, 458 A 3, 458 A 4, 458 A 5, 458 A 6, 458 A 7, 458 A 8, 458 A 9, 458 A 10, 458 A 11

WHITING M'F'G CO., SILVERSMITHS, BROADWAY & 18TH ST., NEW YORK.

TRADE MARK — STERLING

(COPYRIGHTED.) (HALF ACTUAL SIZE.)

WHITING M'F'G CO., SILVERSMITHS, BROADWAY & 18TH ST., NEW YORK.

TRADE MARK STERLING

(HALF ACTUAL SIZE.)

462 A 1 · 462 A 2 · 462 A 3 · 462 A 4 · 462 A 5 · 462 A 6 · 462 A 11 · 462 A 10 · 462 A 9 · 462 A 8

462 A 15 · 462 A 21 · 462 A 20

462 A 7 · 462 A 18 · 462 A 17 · 462 A 19 · 462 A 16 · 462 A 14 · 462 A 13 · 462 A 12

WHITING M'F'G CO.,
SILVERSMITHS,

BROADWAY & 18TH ST.,
NEW YORK.

(HALF ACTUAL SIZE.)

464 A 1 · 464 A 2 · 464 A 3 · 464 A 4 · 464 A 5 · 464 A 6 · 464 A 7 · 464 A 8 · 464 A 9 · 464 A 10 · 464 A 11 · 464 A 12 · 464 A 13

WHITING M'F'G CO., BROADWAY & 18TH ST.,
SILVERSMITHS, NEW YORK.

TRADE MARK STERLING

(COPYRIGHTED.) (HALF ACTUAL SIZE.)

(COPYRIGHTED.) WHITING M'F'G CO., SILVERSMITHS, BROADWAY & 18TH ST., NEW YORK. (HALF ACTUAL SIZE.)

468 A 1

468 A 2

468 A 3

468 A 4

468 A 5

468 A 6

468 A 7

468 A 8

468 A 9

468 A 10

468 A 11

468 A 12

468 A 13

468 A 14

468 A 15

468 A 16

468 A 17

468 A 18

WHITING M'F'G CO., BROADWAY & 18TH ST.,

SILVERSMITHS, NEW YORK.

(HALF ACTUAL SIZE.)

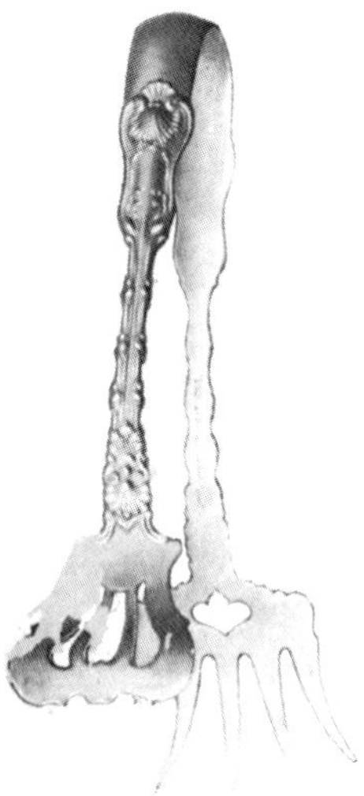

469 A 1

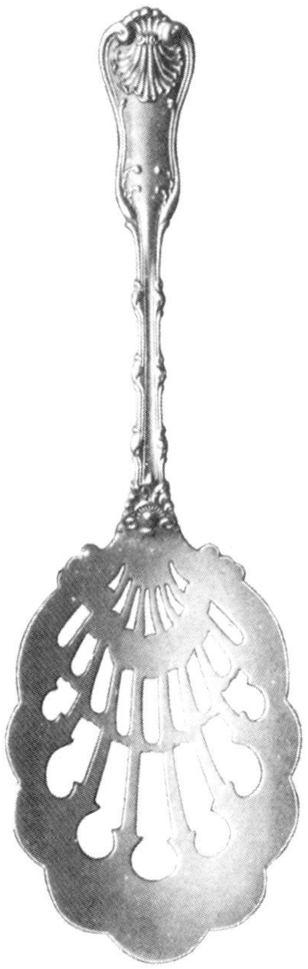

469 A 2

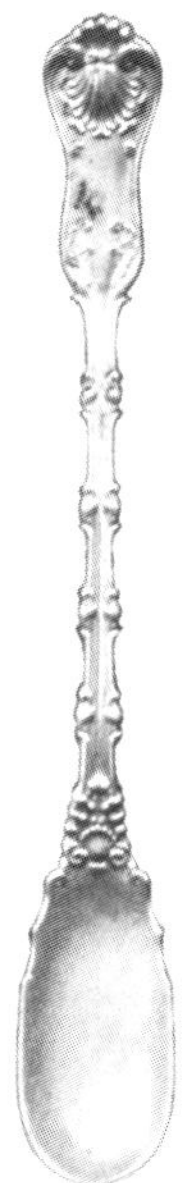

469 A 3

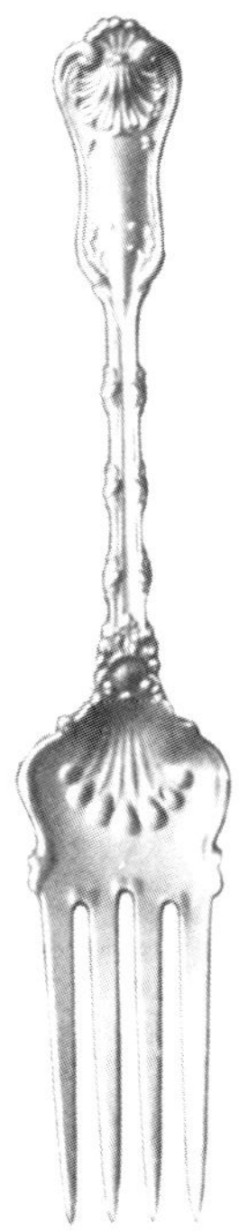

469 A 4

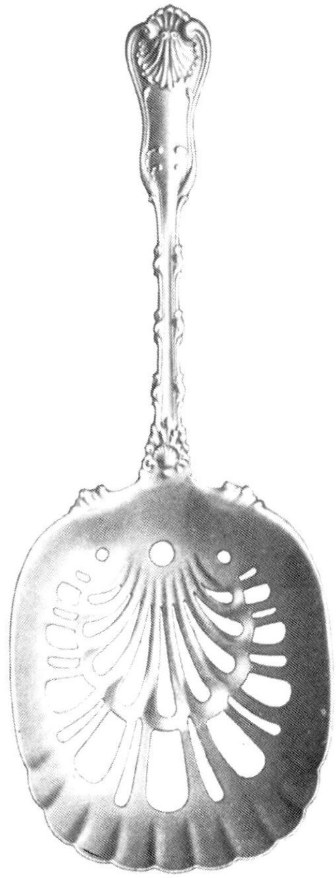

469 A 5

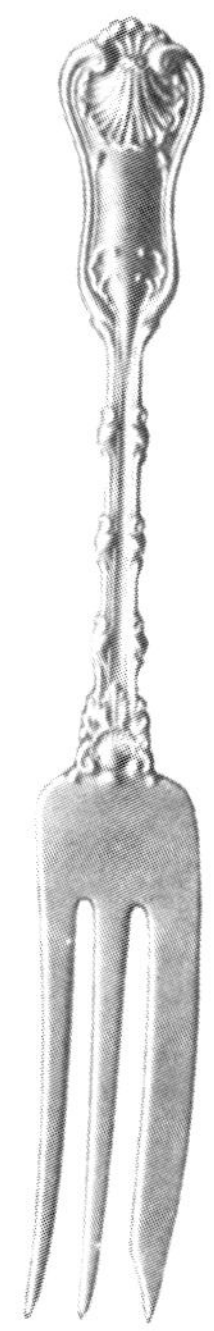

469 A 6

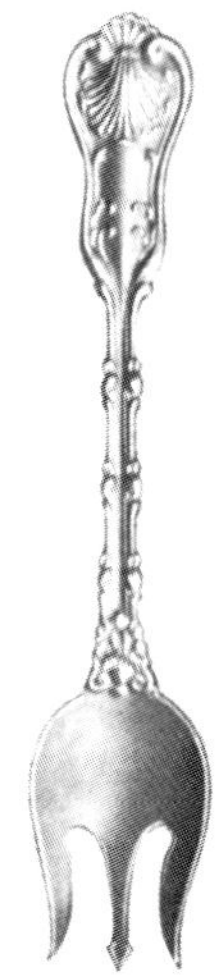

469 A 7

469

(COPYRIGHTED.)

WHITING M'F'G CO., BROADWAY & 18TH ST.,

SILVERSMITHS, NEW YORK.

(HALF ACTUAL SIZE.)

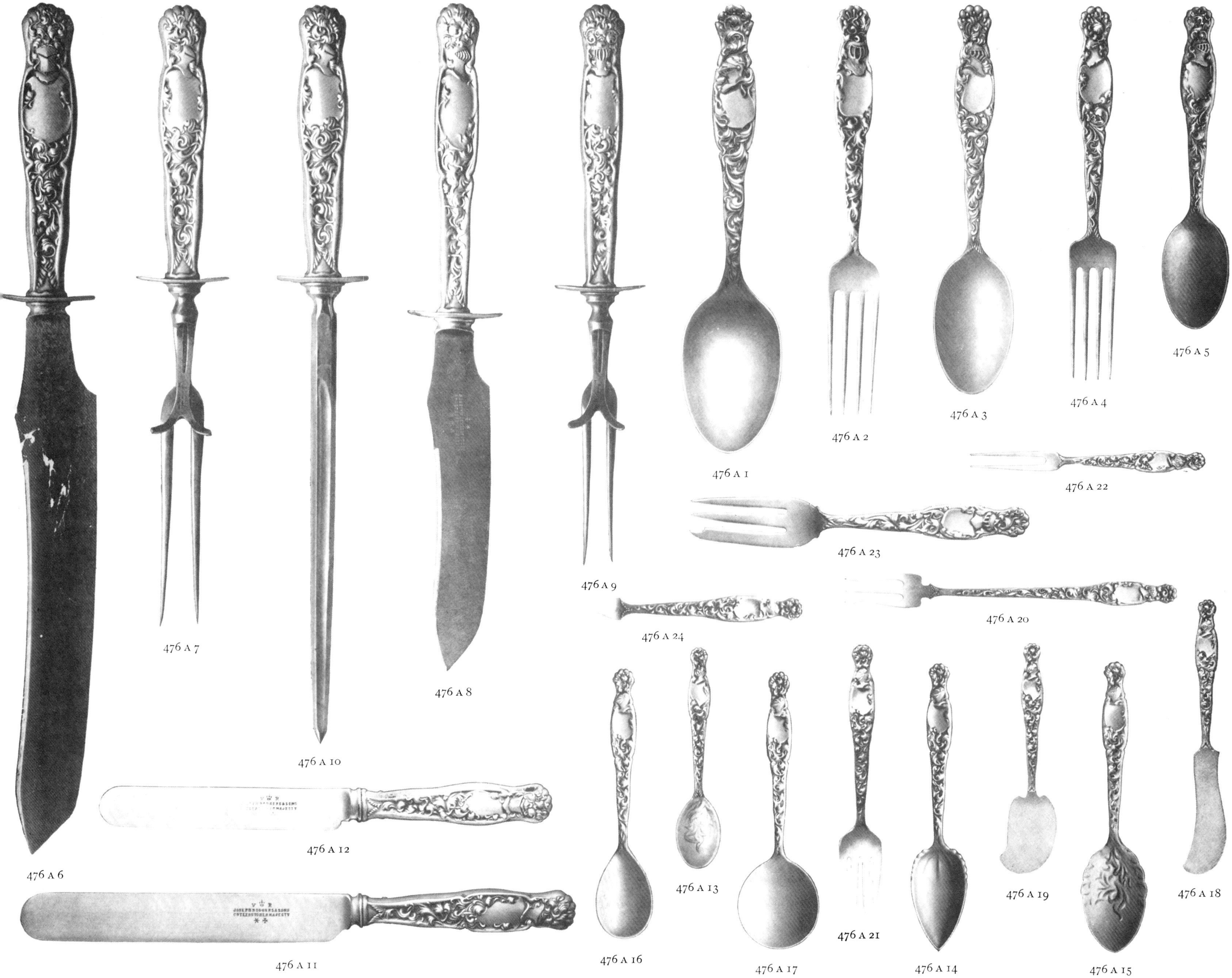

WHITING M'F'G CO., BROADWAY & 18TH ST.,

(COPYRIGHTED.) SILVERSMITHS, NEW YORK. (HALF ACTUAL SIZE.)

478 A 1 — 478 A 2 — 478 A 3 — 478 A 4 — 478 A 6 — 478 A 5 — 478 A 8 — 478 A 9 — 478 A 7

478 A 12 — 478 A 11 — 478 A 10 — 478 A 13 — 478 A 14 — 478 A 15 — 478 A 16 — 478 A 17 — 478 A 19 — 478 A 18

WHITING M'F'G CO., SILVERSMITHS, BROADWAY & 18TH ST., NEW YORK.

TRADE MARK STERLING

(HALF ACTUAL SIZE.)

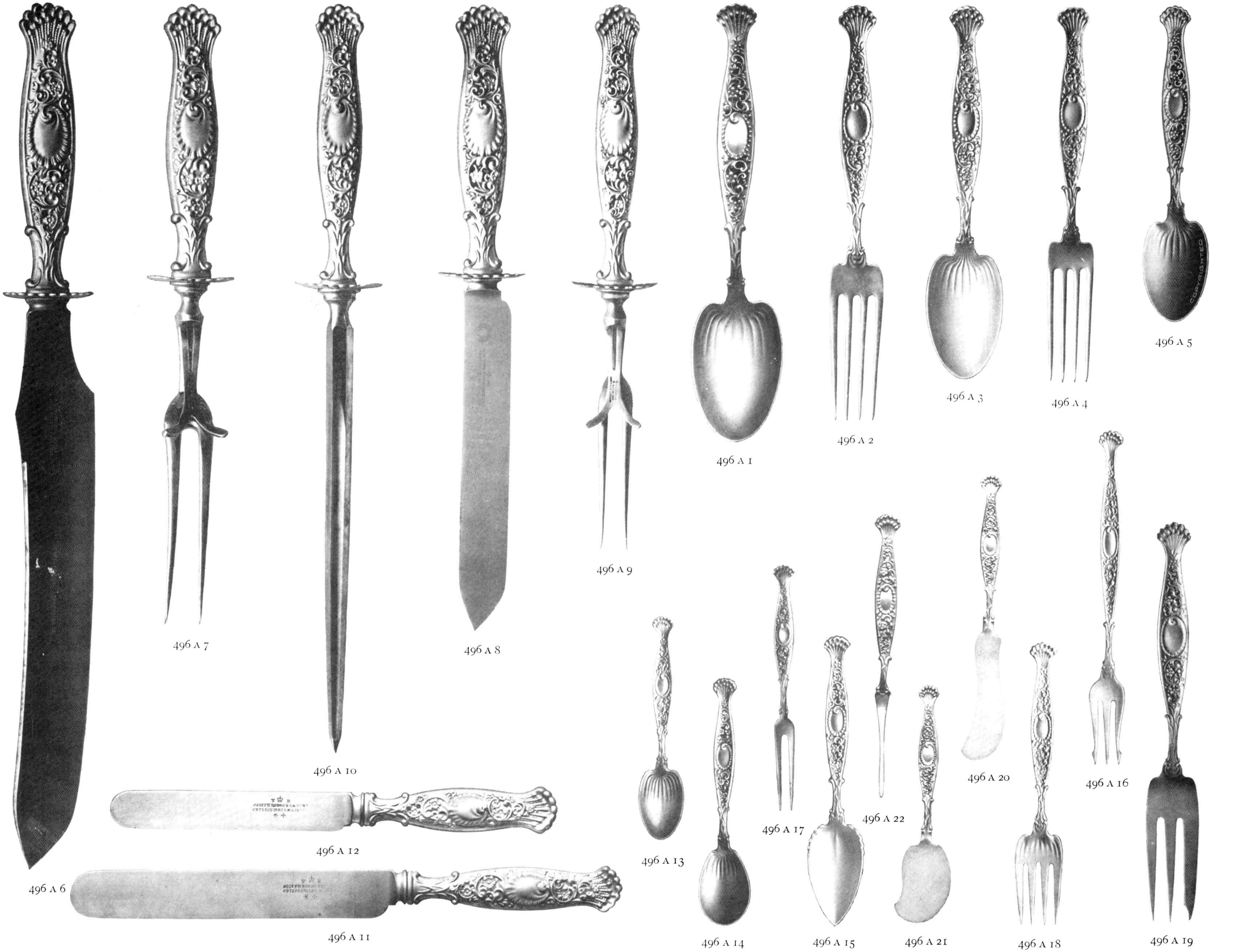

WHITING M'F'G CO., SILVERSMITHS, BROADWAY & 18TH ST., NEW YORK.

(COPYRIGHTED.) (HALF ACTUAL SIZE.)

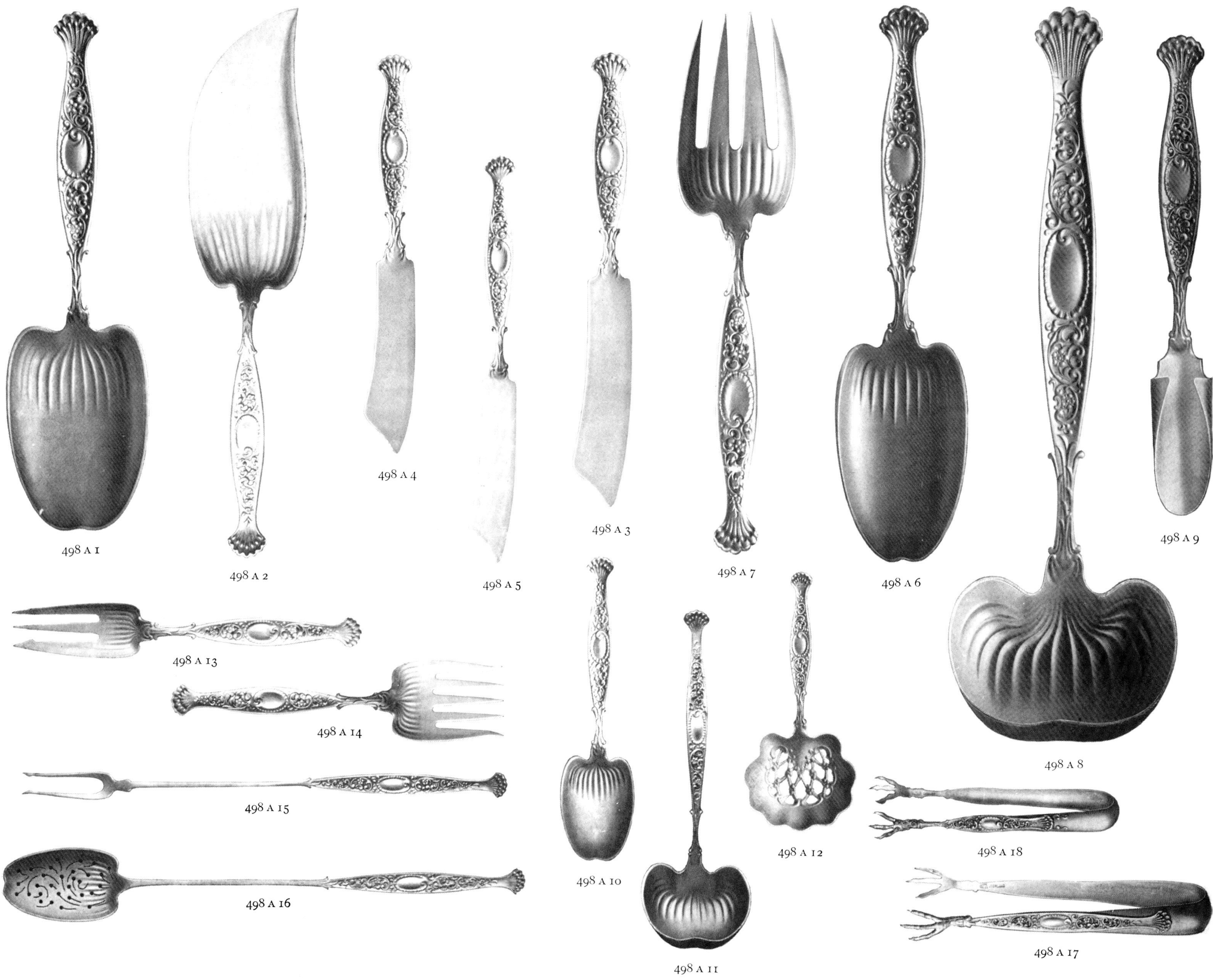

498 A 1 · 498 A 2 · 498 A 4 · 498 A 5 · 498 A 3 · 498 A 7 · 498 A 6 · 498 A 9 · 498 A 8 · 498 A 13 · 498 A 14 · 498 A 15 · 498 A 16 · 498 A 10 · 498 A 11 · 498 A 12 · 498 A 18 · 498 A 17

WHITING M'F'G CO., SILVERSMITHS, BROADWAY & 18TH ST., NEW YORK.

TRADE MARK STERLING

(COPYRIGHTED.) (HALF ACTUAL SIZE.)

"BEAD"

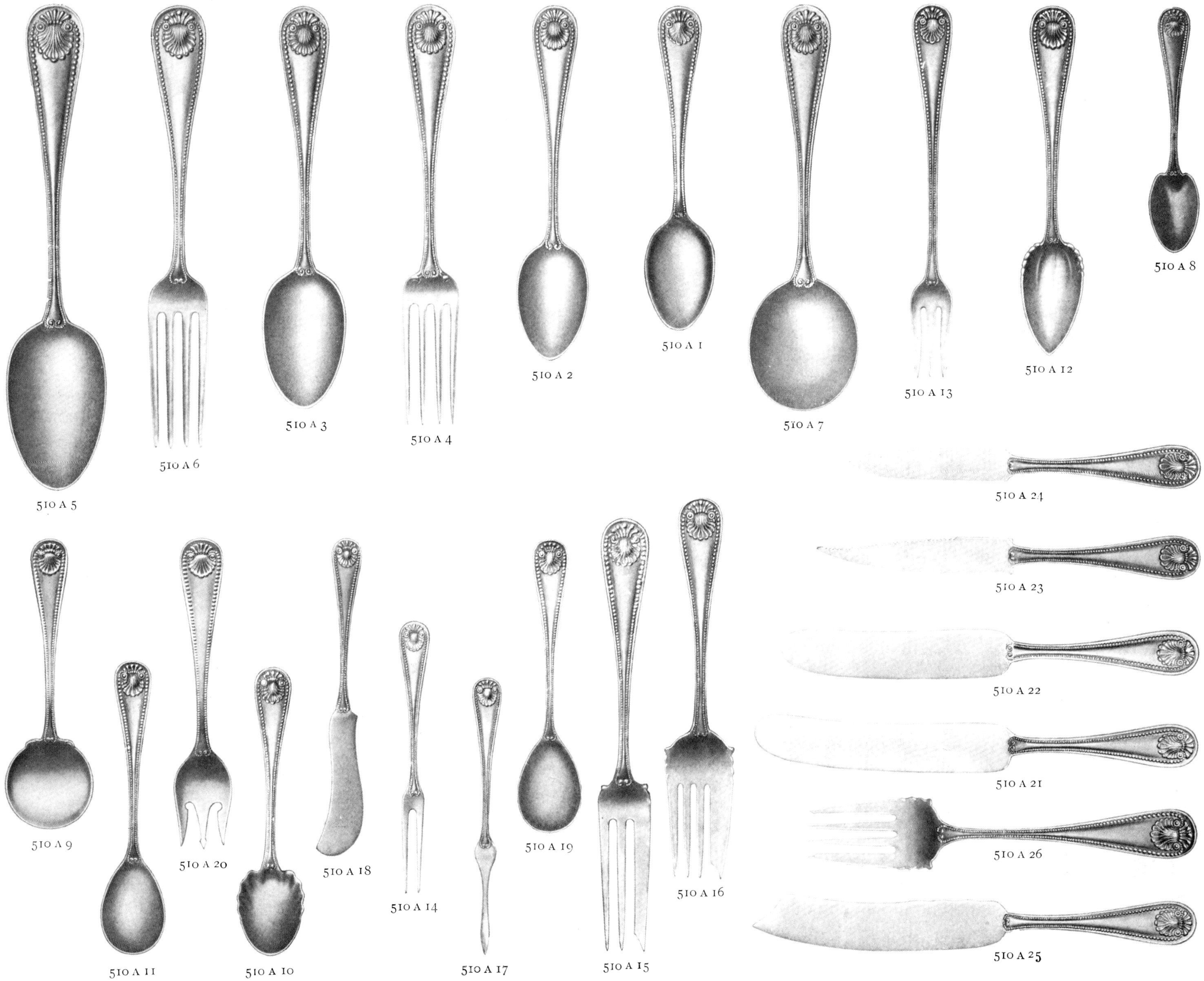

(Copyrighted.) WHITING M'F'G CO., SILVERSMITHS, BROADWAY & 18TH ST., NEW YORK. (Half Actual Size.)

"BEAD"

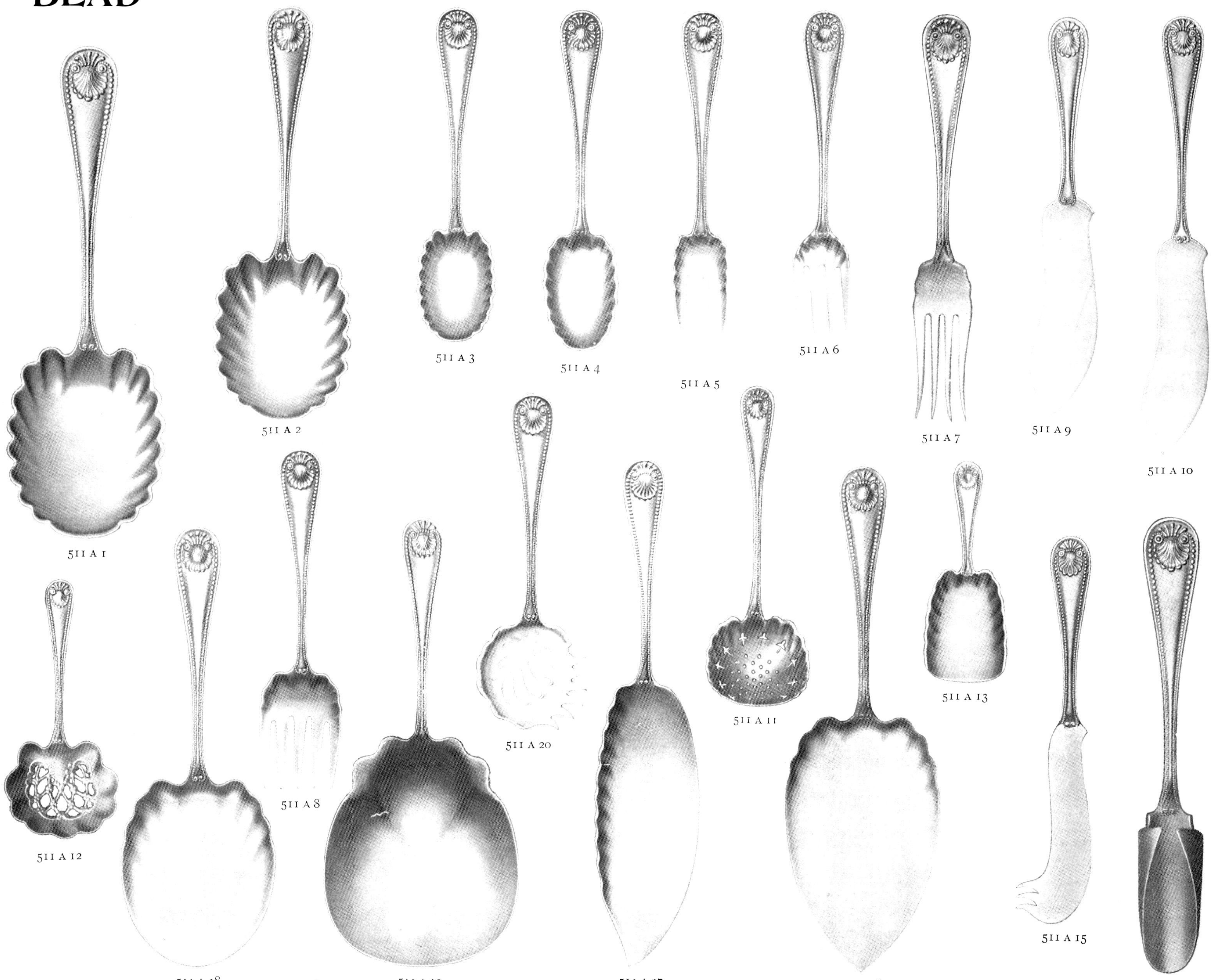

WHITING M'F'G CO., SILVERSMITHS, BROADWAY & 18TH ST., NEW YORK.

(HALF ACTUAL SIZE.)

"BEAD"

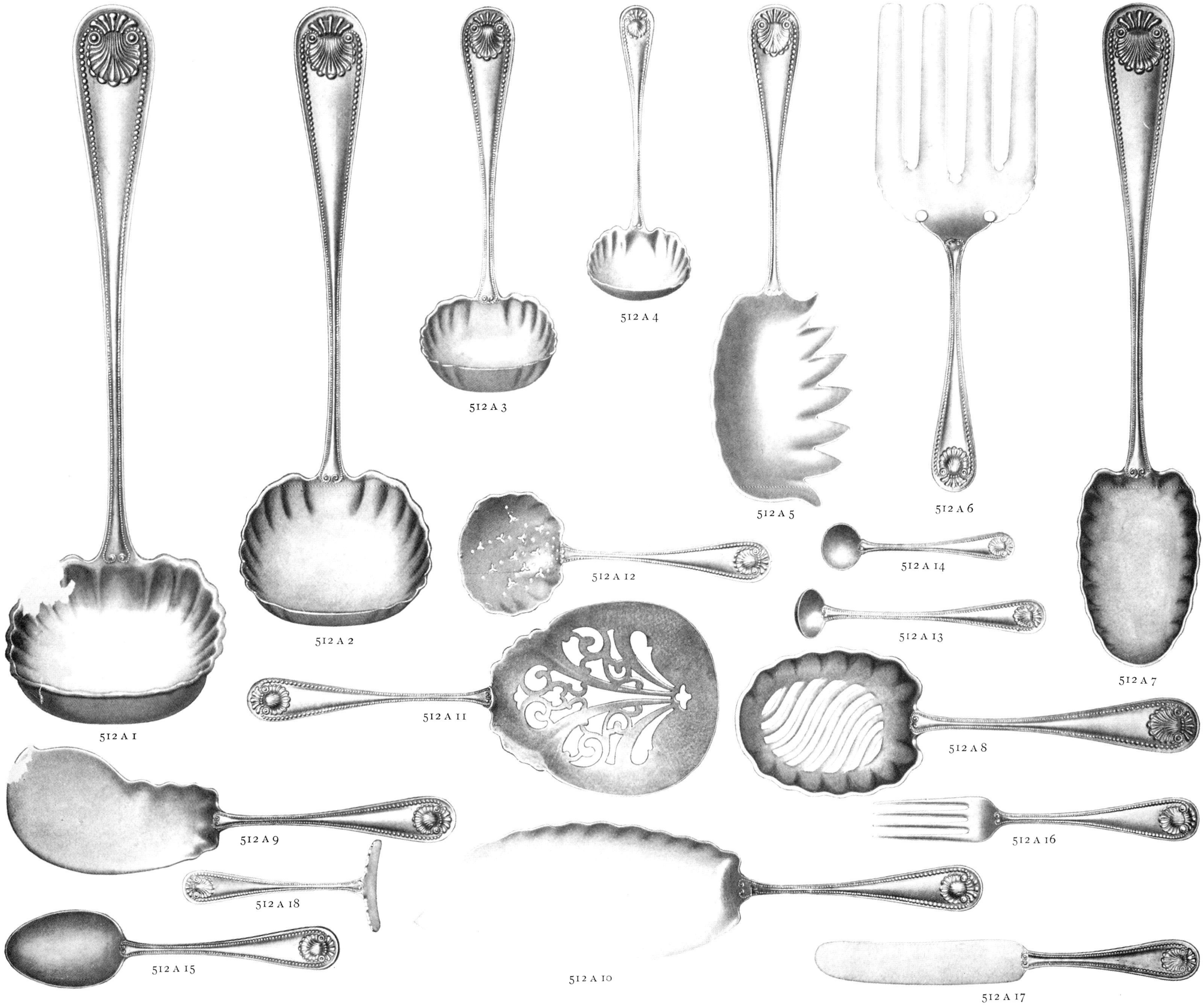

512 A 1, 512 A 2, 512 A 3, 512 A 4, 512 A 5, 512 A 6, 512 A 7, 512 A 8, 512 A 9, 512 A 10, 512 A 11, 512 A 12, 512 A 13, 512 A 14, 512 A 15, 512 A 16, 512 A 17, 512 A 18

WHITING M'F'G CO., BROADWAY & 18TH ST.,

SILVERSMITHS, NEW YORK.

(HALF ACTUAL SIZE.)

"BEAD"

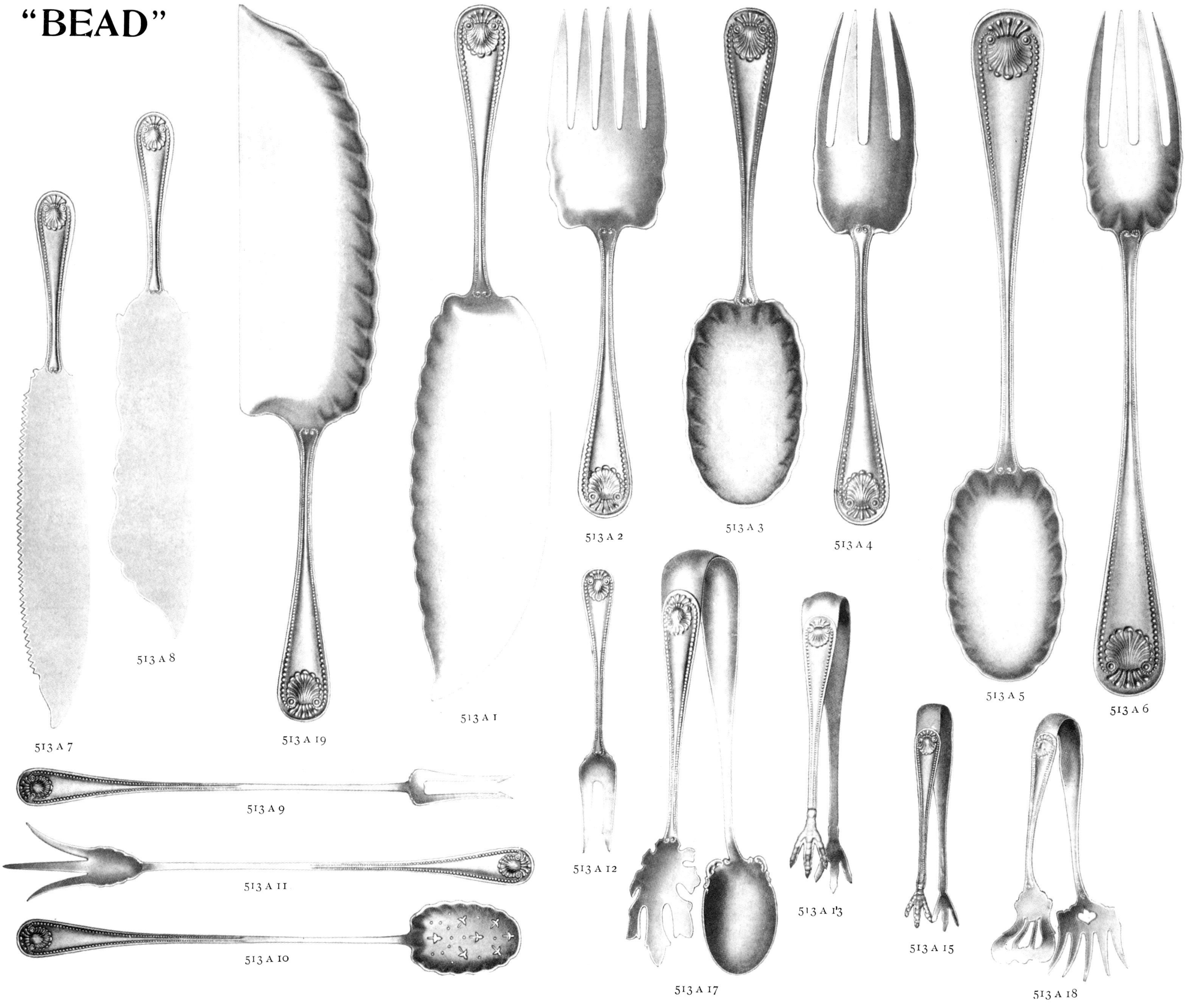

513 A 7 — 513 A 8 — 513 A 19 — 513 A 1 — 513 A 2 — 513 A 3 — 513 A 4 — 513 A 5 — 513 A 6

513 A 9 — 513 A 11 — 513 A 10 — 513 A 12 — 513 A 17 — 513 A 13 — 513 A 15 — 513 A 18

WHITING M'F'G CO., SILVERSMITHS, BROADWAY & 18TH ST., NEW YORK.

TRADE MARK STERLING

(HALF ACTUAL SIZE.)

"LIGHT OVAL TWIST"

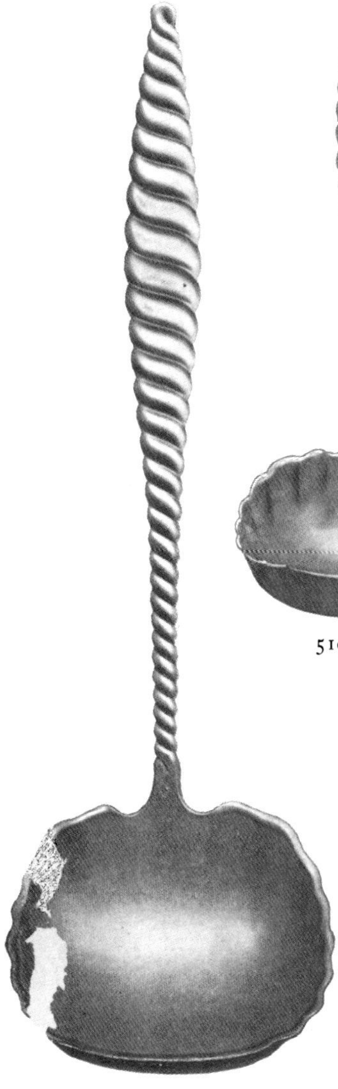

519 A 1

519 A 2

519 A 3

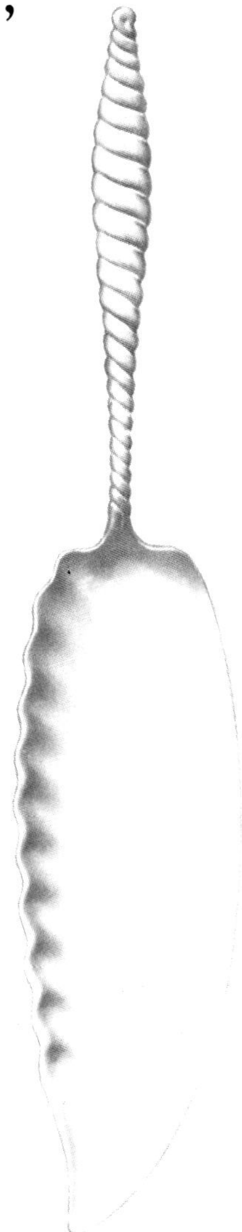

519 A 4

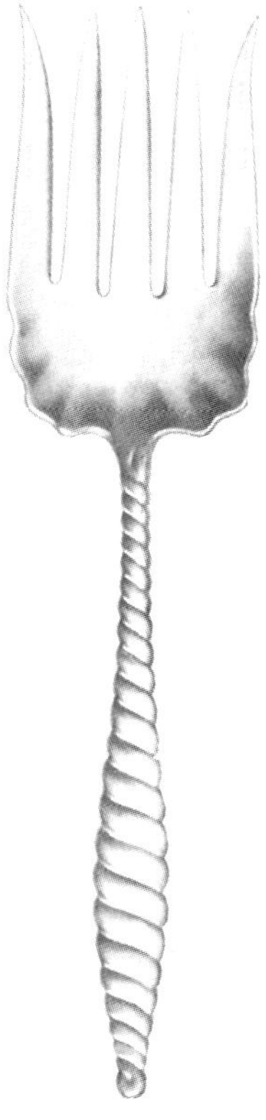

519 A 5

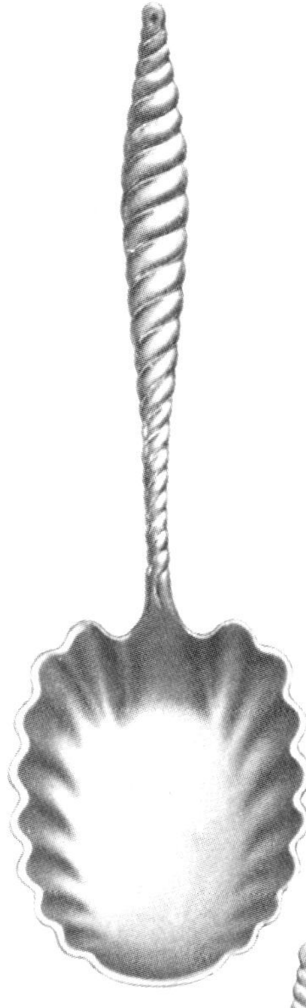

519 A 6

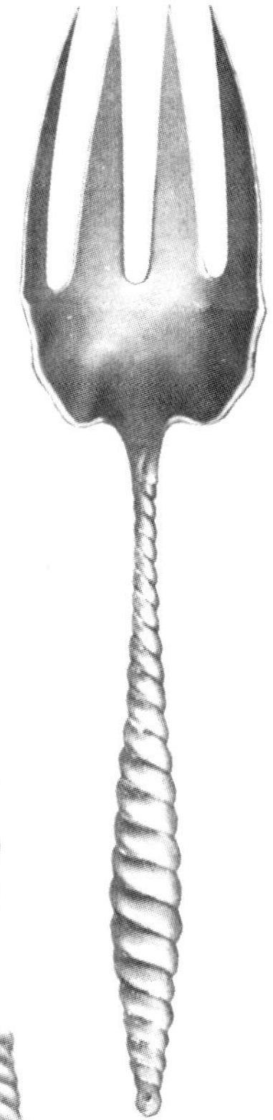

519 A 7

519 A 8

519 A 9

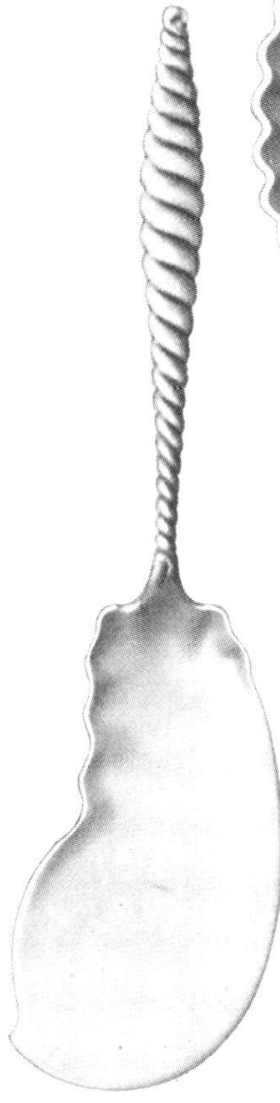

519 A 10

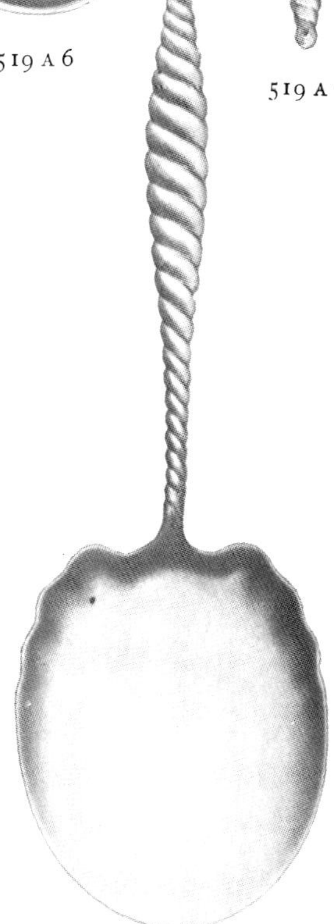

519 A 11

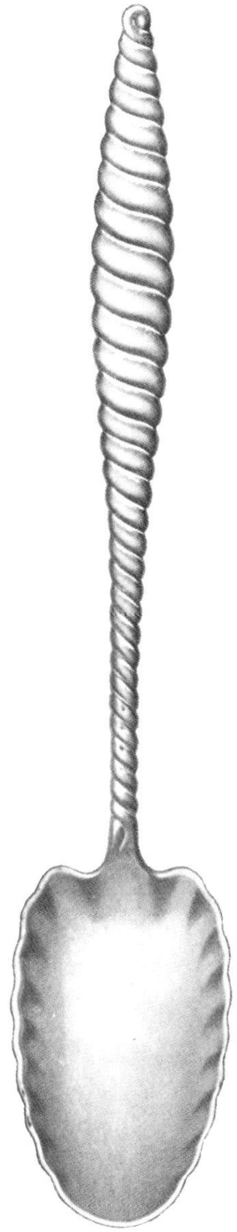

519 A 12

WHITING M'F'G CO., SILVERSMITHS, BROADWAY & 18TH ST., NEW YORK.

(HALF ACTUAL SIZE.)

"LIGHT OVAL TWIST"

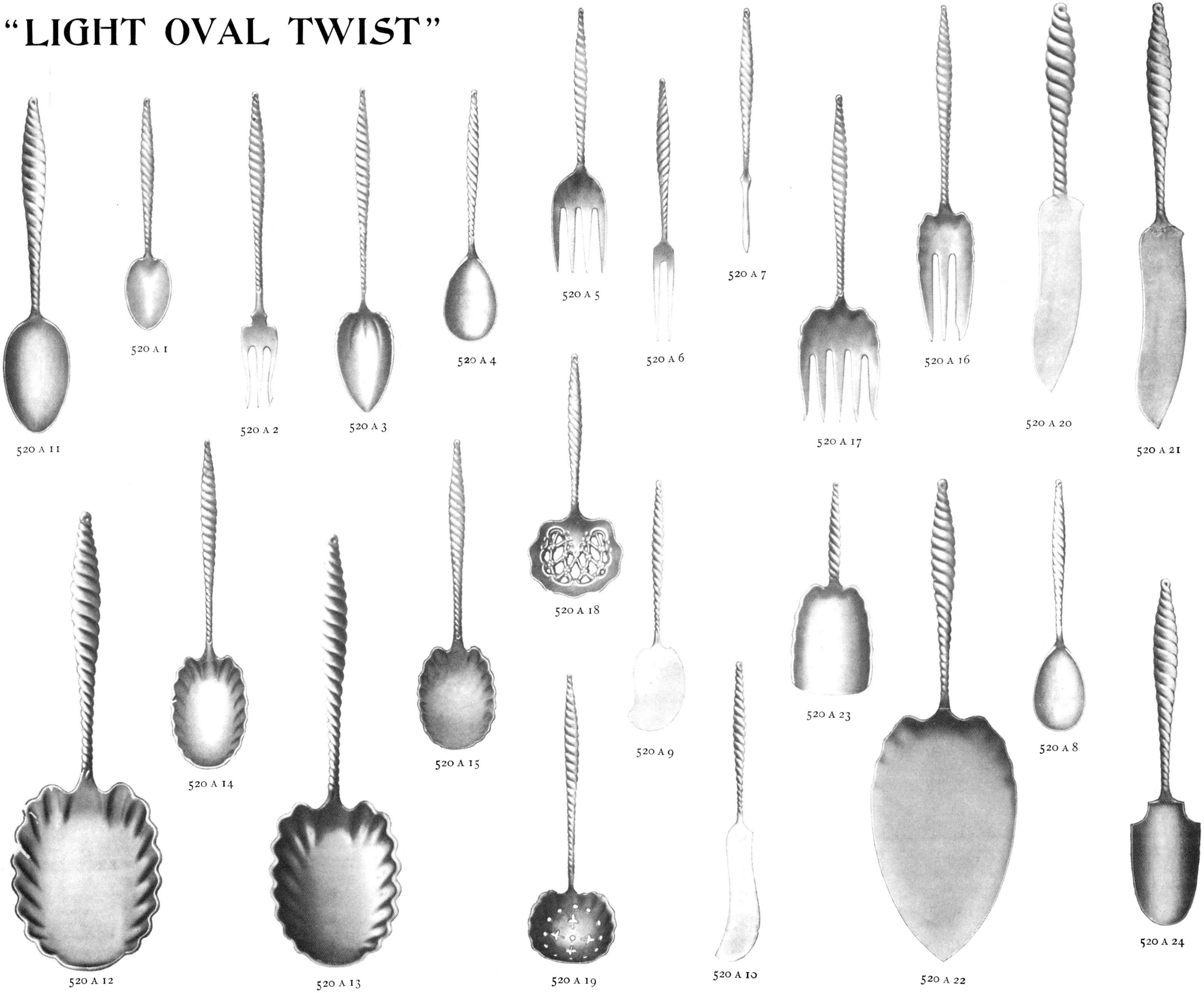

WHITING M'F'G CO., SILVERSMITHS, BROADWAY & 18TH ST., NEW YORK.

TRADE MARK STERLING

(COPYRIGHTED.)

(HALF ACTUAL SIZE.

"LIGHT OVAL TWIST"

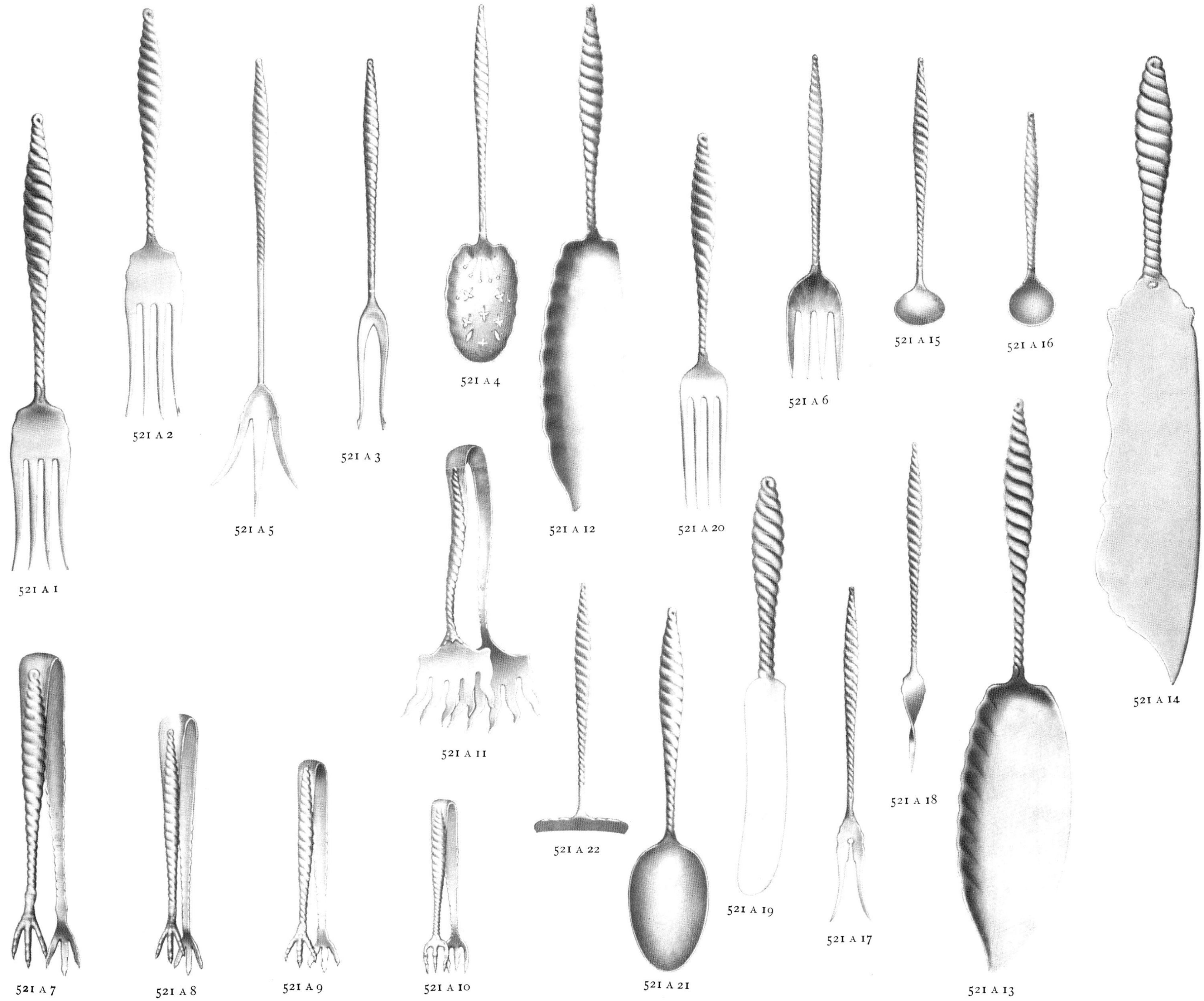

WHITING M'F'G CO., SILVERSMITHS, BROADWAY & 18TH ST., NEW YORK.

(HALF ACTUAL SIZE.)

548 A 1

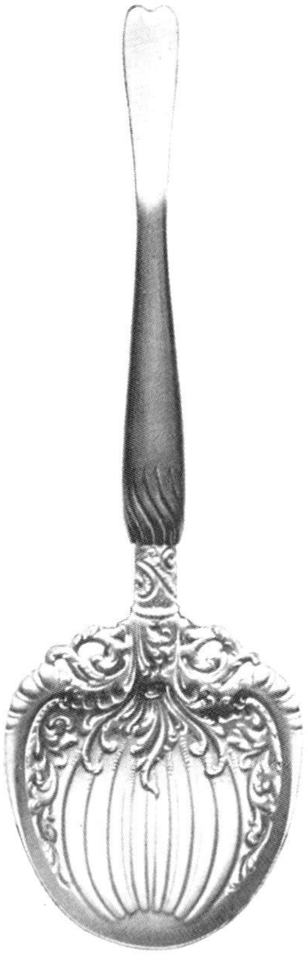

548 A 2

548 A 12

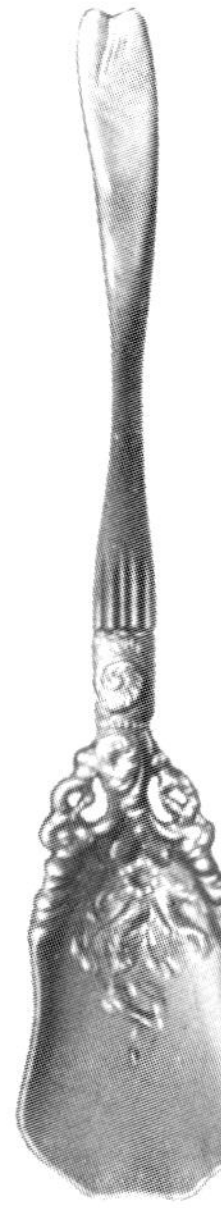

548 A 3

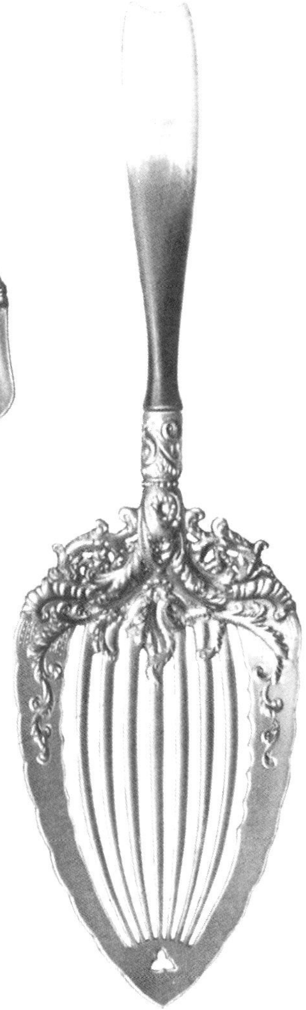

548 A 13

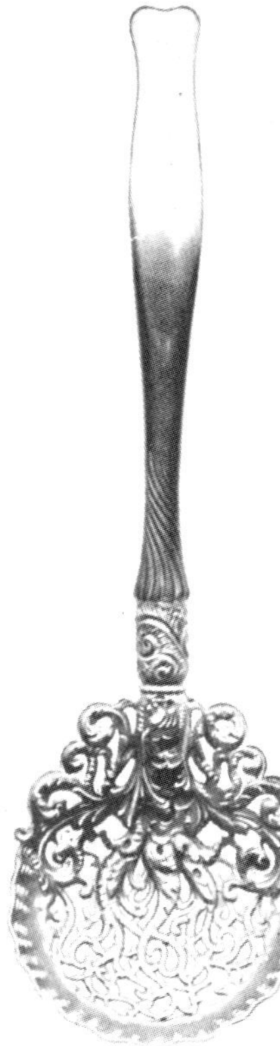

548 A 4

548 A 14

548 A 5

548 A 9

548 A 10

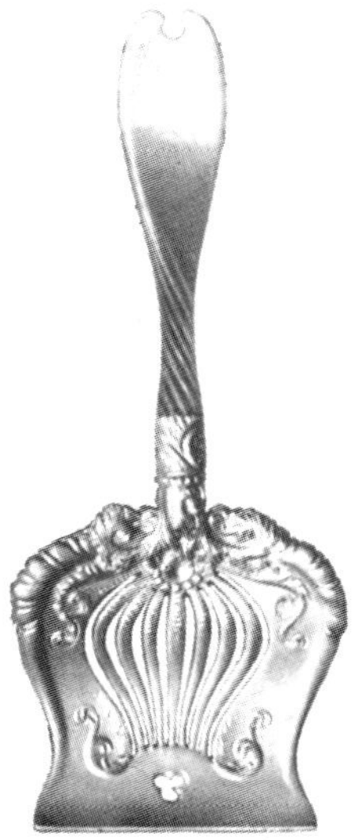

548 A 11

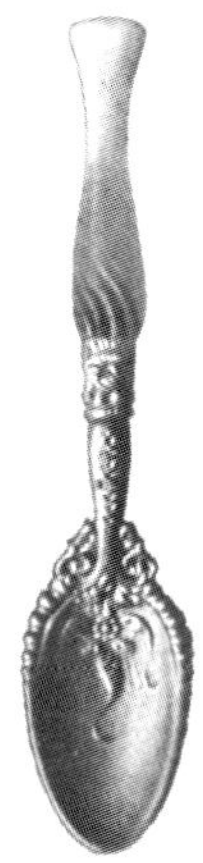

548 A 6

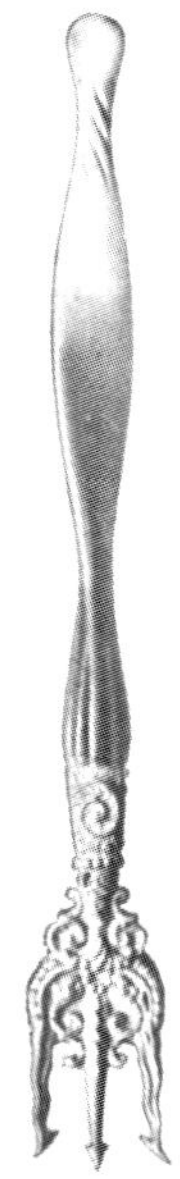

548 A 7

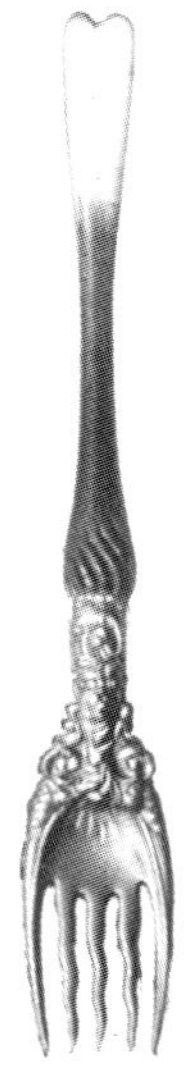

548 A 8

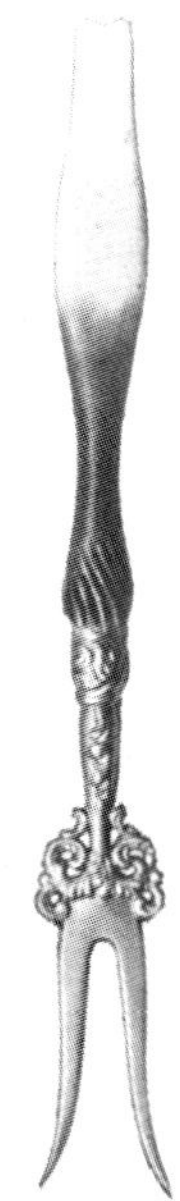

548 A 15

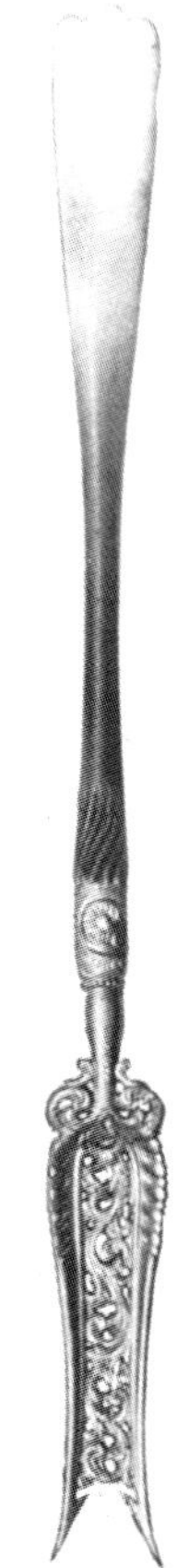

548 A 16

WHITING M'F'G CO., SILVERSMITHS, BROADWAY & 18TH ST., NEW YORK.

TRADE MARK STERLING

(HALF ACTUAL SIZE.)

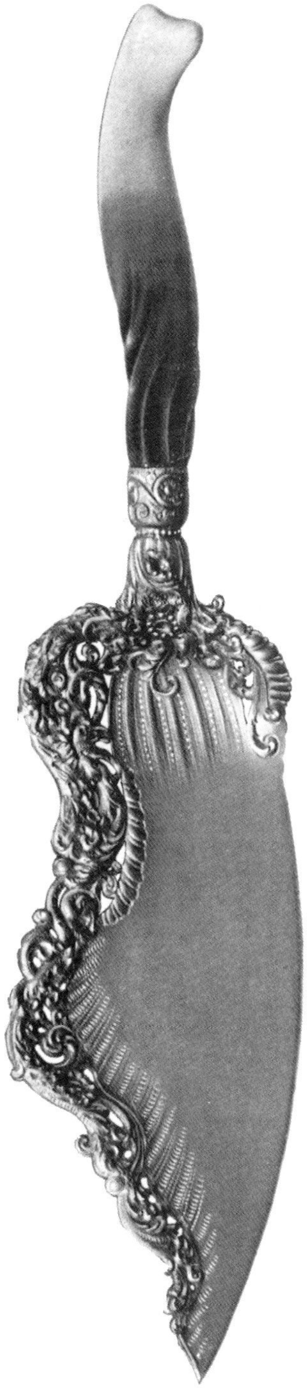

549 A 1

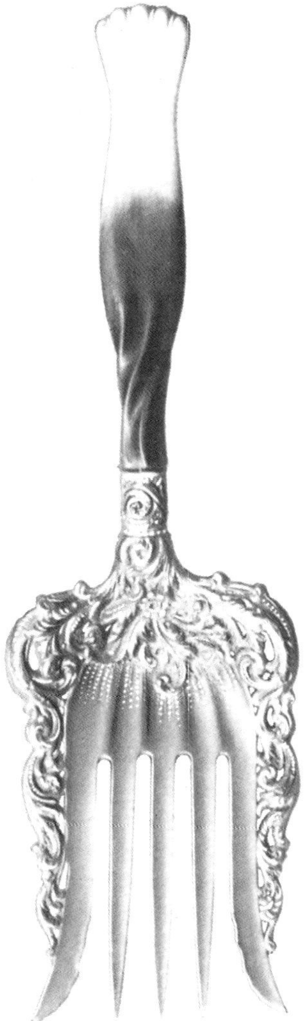

549 A 2

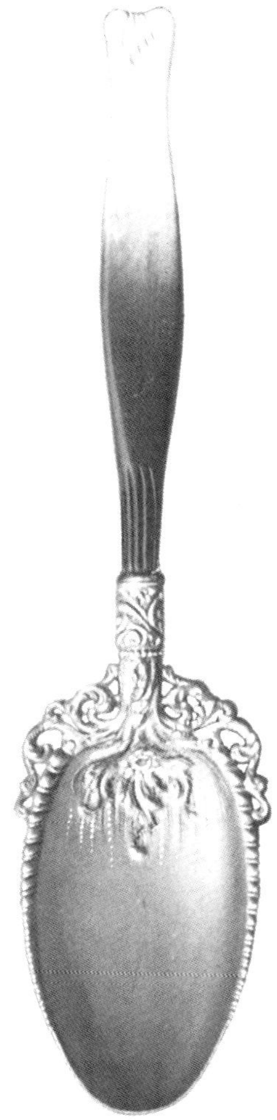

549 A 3

549 A 4

549 A 5

549 A 6

549 A 7

549 A 8

WHITING M'F'G CO.,
SILVERSMITHS,

BROADWAY & 18TH ST.,
NEW YORK.

(HALF ACTUAL SIZE.)

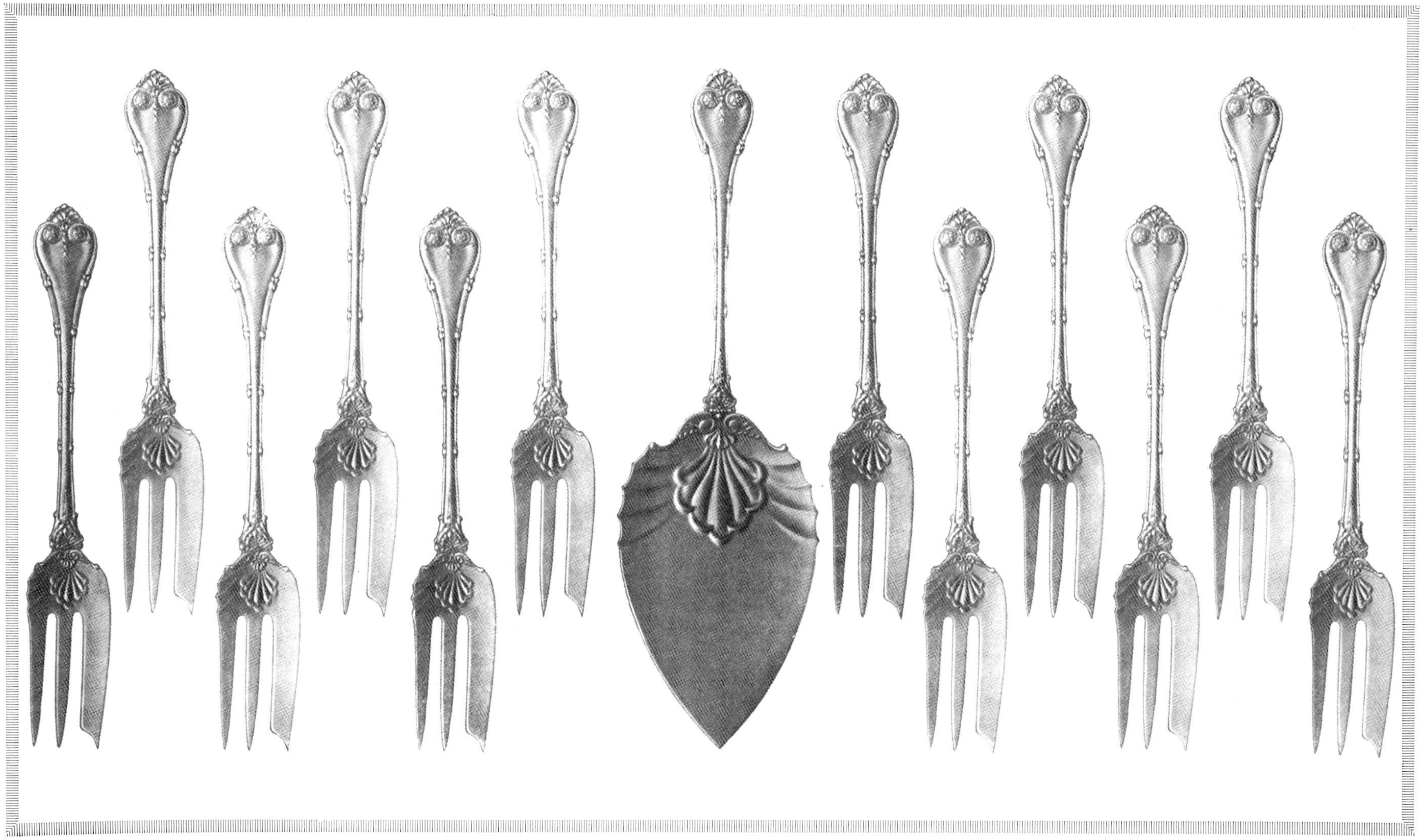

601 A 1

(Copyrighted.) WHITING M'F'G CO., SILVERSMITHS, BROADWAY & 18TH ST., NEW YORK. (Half Actual Size.)

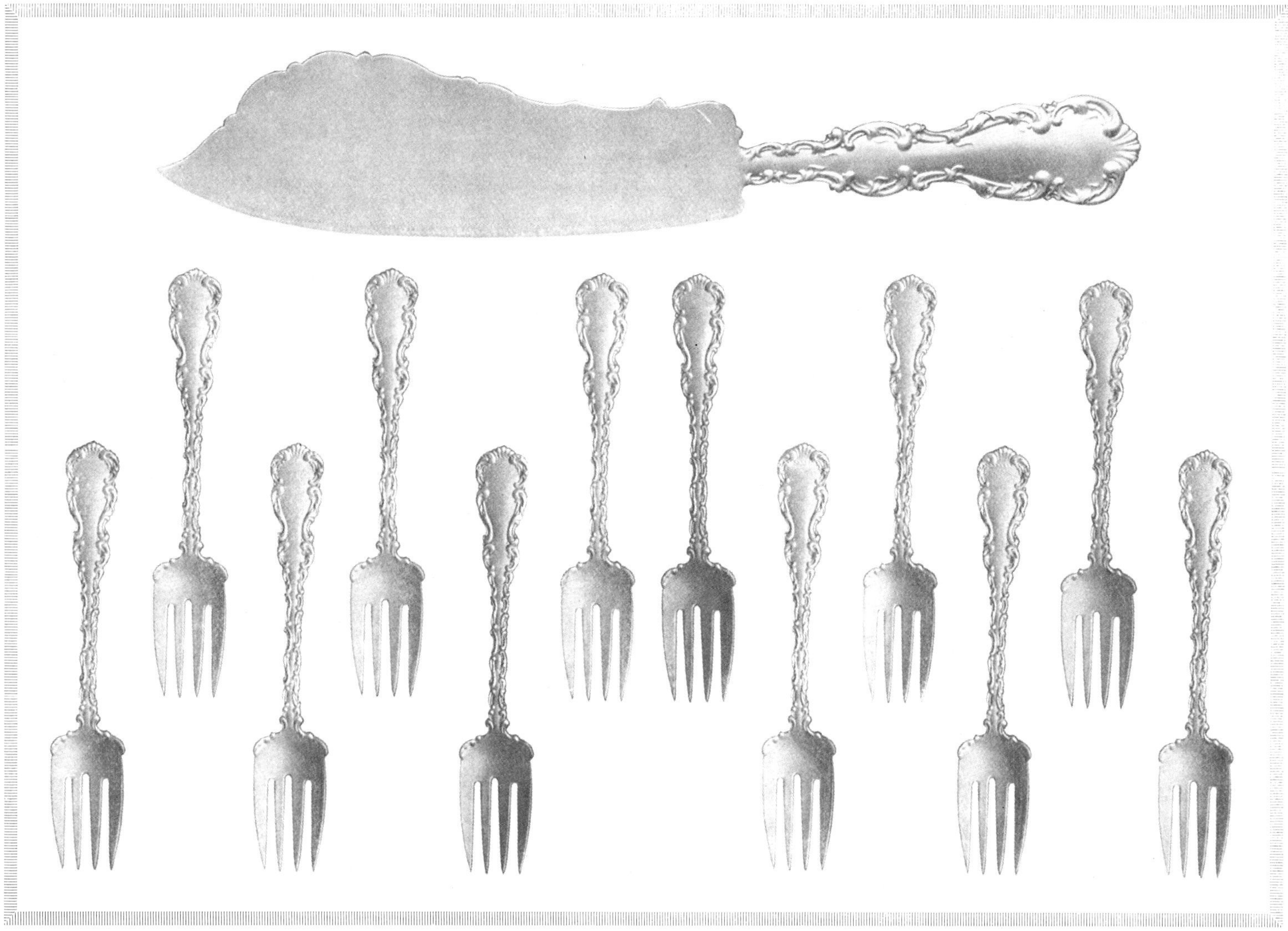

609 A 1

658 A 1

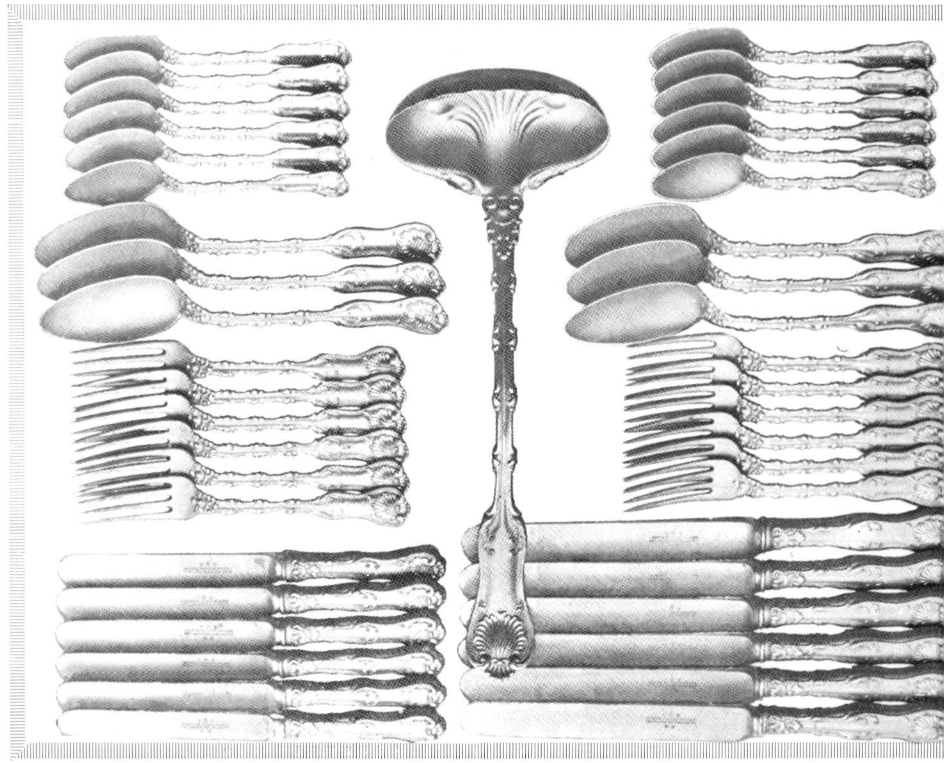

658 A 3

658 A 2

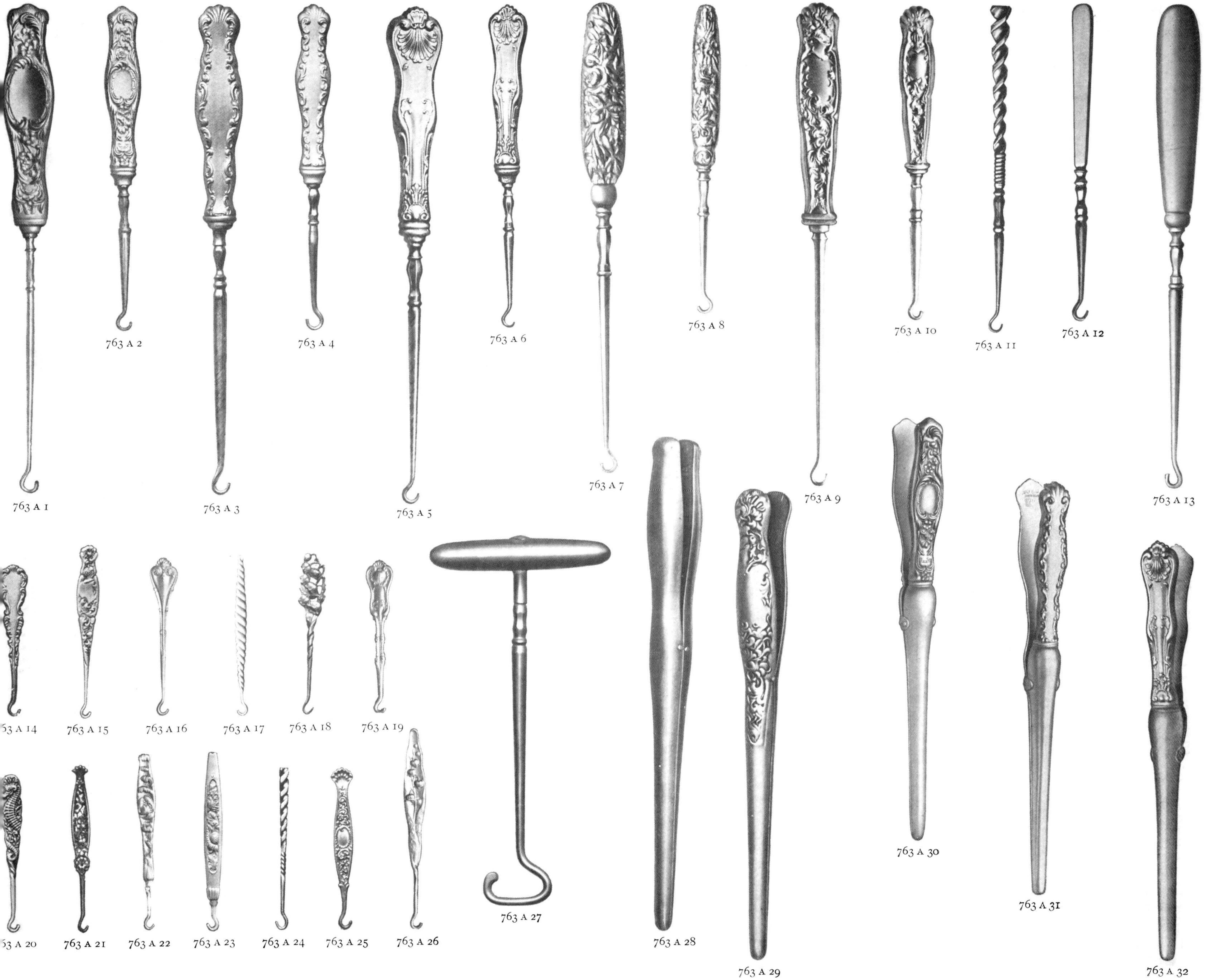

763 A 2 763 A 4 763 A 6 763 A 8 763 A 10 763 A 11 763 A 12

763 A 1 763 A 3 763 A 5 763 A 7 763 A 9 763 A 13

63 A 14 763 A 15 763 A 16 763 A 17 763 A 18 763 A 19

63 A 20 763 A 21 763 A 22 763 A 23 763 A 24 763 A 25 763 A 26 763 A 27

763 A 28 763 A 29 763 A 30 763 A 31 763 A 32

WHITING M'F'G CO., SILVERSMITHS, BROADWAY & 18TH ST., NEW YORK.

TRADE MARK STERLING

(HALF ACTUAL SIZE.)

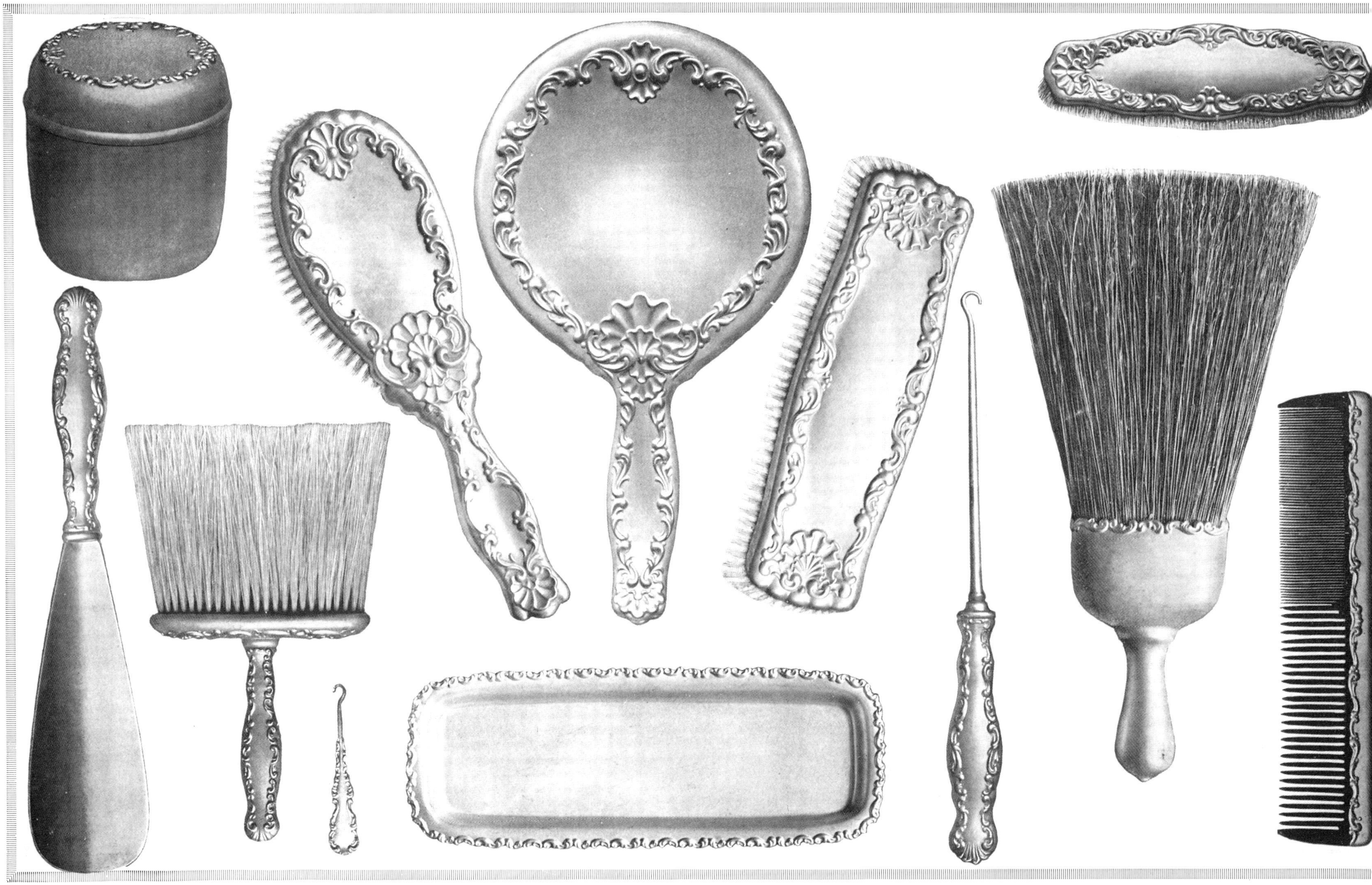

775 A 117

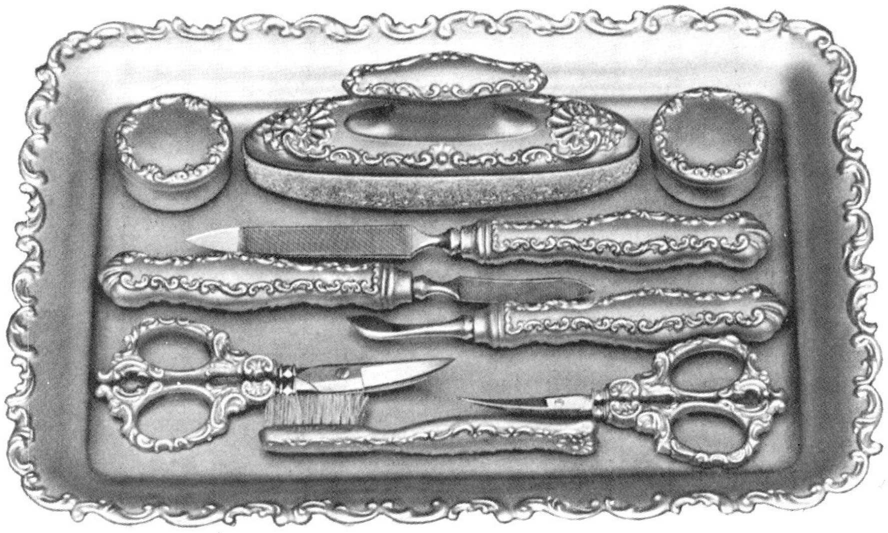

775 A 214

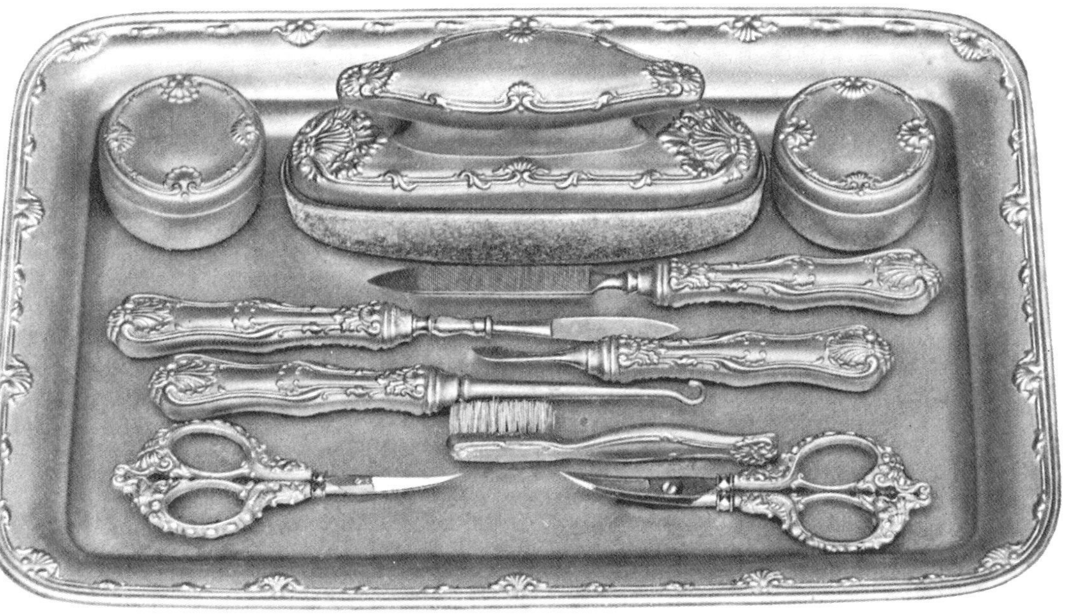

775 A 215

775 A 266

(HALF ACTUAL SIZE.)

775 A 319

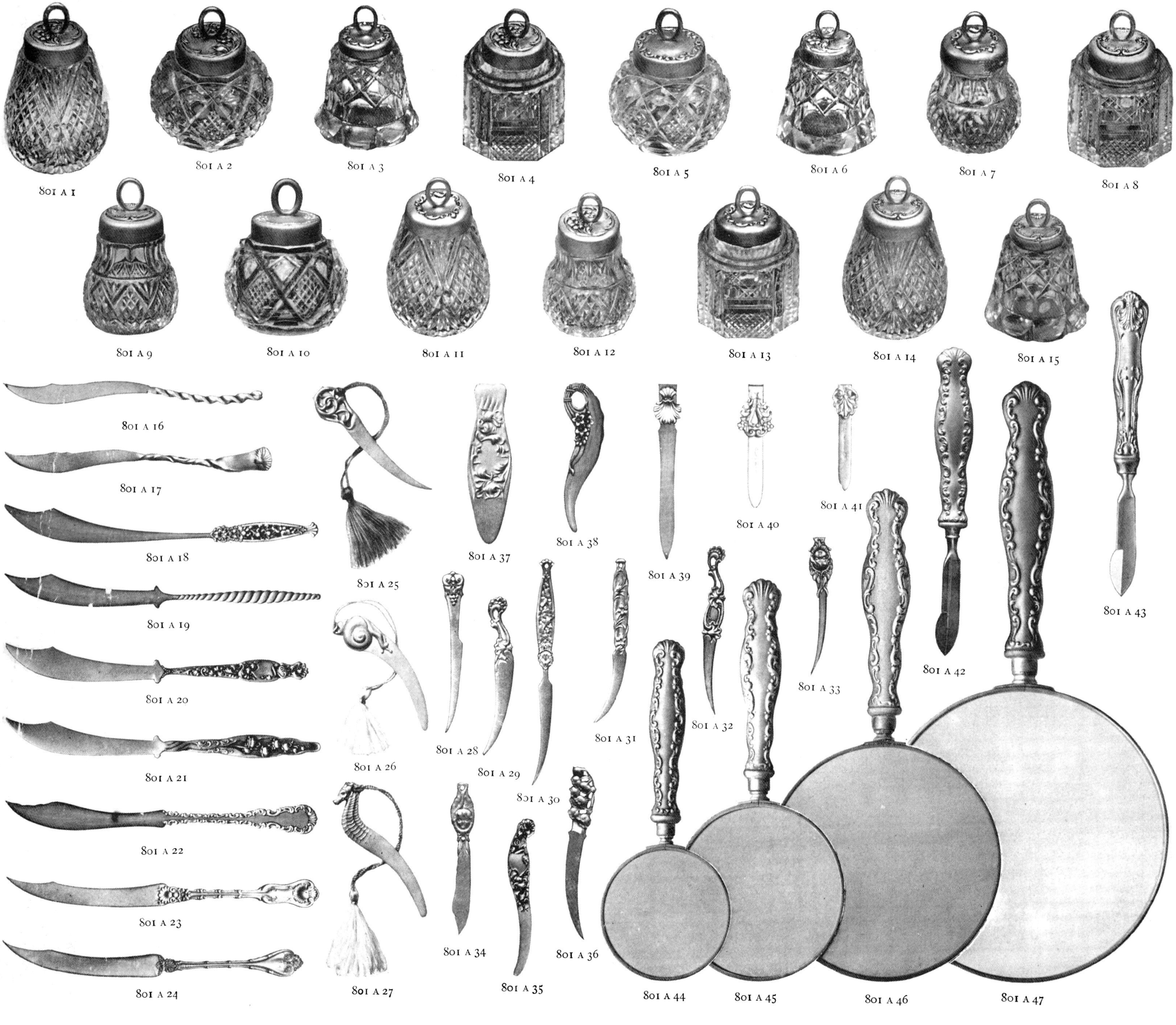
801 A 1
801 A 2
801 A 3
801 A 4
801 A 5
801 A 6
801 A 7
801 A 8
801 A 9
801 A 10
801 A 11
801 A 12
801 A 13
801 A 14
801 A 15
801 A 16
801 A 17
801 A 18
801 A 19
801 A 20
801 A 21
801 A 22
801 A 23
801 A 24
801 A 25
801 A 26
801 A 27
801 A 28
801 A 29
801 A 30
801 A 31
801 A 32
801 A 33
801 A 34
801 A 35
801 A 36
801 A 37
801 A 38
801 A 39
801 A 40
801 A 41
801 A 42
801 A 43
801 A 44
801 A 45
801 A 46
801 A 47

WHITING M'F'G CO., SILVERSMITHS, BROADWAY & 18TH ST., NEW YORK.

Trade Mark Sterling

(Copyrighted.)

(Half Actual Size.)

Index to the Numbered Items

The original catalogue was believed to have been accompanied by a price list, which would have corresponded, to the numbers listed in this catalogue. Unfortunately, this price list did not survive over the years. Today, we are left with the task of reconstructing an index to these numbered pieces. Using other catalogues by Whiting and other manufacturers we have attempted to create an accurate list. Although we have been able to identify most pieces, there are some for which we were unable to identify any further than "Fork". Hopefully, over time, these few remaining pieces will be identified by the discovery of additional Whiting catalogues.

No.	Item
2 A 40	Heraldic: Butter Spreaders & Butter Pats
0 A 9	Louis XV: Teaspoons, Sugar Tongs & 3-Piece Coffee Set
0 A 17	Imperial Queen: Teaspoons, Sugar Tongs, 3 Piece Coffee Set & Tray
245 A 1	Heraldic: Tea Strainer
245 A 2	Lily of The Valley: Tea Strainer
245 A 3	Louis XV: Tea Strainer
245 A 4	Hyperion: Tea Strainer
245 A 5	Imperial Queen: Tea Strainer
245 A 6	Oval Twist: Tea Strainer
245 A 7	Empire: Tea Strainer
245 A 8	Tea Strainer
245 A 9	Tea Strainer
245 A 10	Tea Strainer
245 A 11	Tea Strainer
245 A 12	Tea Strainer
245 A 13	Imperial Queen: Tea Strainer
245 A 14	Ivory: Tea Strainer
245 A 15	Tea Ball
245 A 16	Tea Ball
245 A 17	Tea Ball
245 A 18	Tea Ball
245 A 19	Tea Ball
245 A 20	Tea Ball
245 A 21	Tea Ball
245 A 22	Tea Ball
245 A 23	Tea Ball
245 A 24	Tea Ball Holder
245 A 25	Tea Ball Holder
245 A 26	Tea Ball Holder
248 A 43	Imperial Queen: Caddy Scoop & Tea Ball
248 A 44	Empire: Tea Caddy, Caddy Scoop & Tea Ball
248 A 45	Louis XV: Teaspoons, Sugar Tongs, Tea Caddy, Caddy Scoop, Tea Ball, Creamer & Sugar
248 A 79	Dresden: Berry Spoon, Sugar Sifter, Creamer, Sugar & Bowl
248 A 87	Louis XV: Berry Spoon, Sugar Sifter, Berry Forks, Berry Bowl, Creamer & Sugar
345 A 125	Louis XV: Nut Picks, Nut Spoon, Nut Cracker & Nut Bowl
345 A 149	Louis XV: Ice Cream Slicer, Ice Cream Spoons, Ice Cream Tray & Plates
376 A 1	Imperial Queen: Child's Knife
376 A 2	Imperial Queen: Child's Fork
376 A 3	Imperial Queen: Child's Spoon
376 A 4	Imperial Queen: Child's Knife (Small)
376 A 5	Imperial Queen: Child's Fork (Small)
376 A 6	Imperial Queen: Child's Spoon (Small)
376 A 7	Louis XV: Child's Knife
376 A 8	Louis XV: Child's Fork
376 A 9	Louis XV: Child's Spoon
376 A 10	Louis XV: Child's Knife (Small)
376 A 11	Louis XV: Child's Fork (Small)
376 A 12	Louis XV: Child's Spoon (Small)
376 A 13	Empire: Child's Knife
376 A 14	Empire: Child's Fork
376 A 15	Empire: Child's Spoon
376 A 16	Louis XV: Spoon
376 A 17	Imperial Queen: Spoon
376 A 18	Empire: Spoon
376 A 19	Empire: Food Pusher
376 A 20	Imperial Queen: Food Pusher
376 A 21	Light Oval Twist: Food Pusher
376 A 22	Heraldic: Food Pusher
376 A 23	Louis XV: Food Pusher
376 A 24	Hyperion: Food Pusher
376 A 25	Baby Rattle
376 A 26	Baby Rattle
376 A 27	Baby Rattle
376 A 28	Baby Rattle
376 A 29	Baby Rattle
376 A 30	Baby Rattle
376 A 31	Baby Rattle
376 A 32	Baby Rattle
376 A 33	Baby Rattle
380 A 1	Louis XV: Child's Knife, Fork and Spoon & Napkin Ring
380 A 2	Empire: Child's Knife, Fork and Spoon, & Child's Cup
380 A 3	Heraldic: Child's Knife, Fork and Spoon, Child's Cup & Napkin Ring
380 A 4	Louis XV: Child's Knife, Fork and Spoon, Child's Cup and Plate & Napkin Ring
380 A 31	Empire: Child's Knife, Fork and Spoon, Child's Cup and Plate, Child's Bowl, Child's Porringer, Napkin Ring & Spoon
422 A 1	Empire: Meat Carving Knife, H.H.
422 A 2	Empire: Meat Carving Fork, H.H.
422 A 3	Empire: Game Carving Knife, H.H.
422 A 4	Empire: Game Carving Fork. H.H.
422 A 5	Empire: Steel, H.H.
422 A 6	Empire: Beefsteak Carving Knife, H.H.
422 A 7	Empire: Beefsteak Carving Fork, H.H.
422 A 8	Empire: Steel, H.H. (Small)
422 A 9	Empire: Table Knife
422 A 10	Empire: Medium Knife
422 A 11	Empire: Dessert Knife
424 A 1	Empire: Table Spoon
424 A 2	Empire: Table Fork
424 A 3	Empire: Dessert Spoon
424 A 4	Empire: Dessert Fork
424 A 5	Empire: Teaspoon
424 A 6	Empire: Teaspoon, Five O'clock
424 A 7	Empire: Soup Spoon
424 A 8	Empire: Coffee Spoon
424 A 9	Empire: Spoon
424 A 10	Empire: Boullion Spoon
424 A 11	Empire: Ice Cream Spoon
424 A 12	Empire: Egg Spoon
424 A 13	Empire: Orange Spoon
424 A 14	Empire: Salt Spoon (Indv.)
424 A 15	Empire: Oyster Fork
424 A 16	Empire: Ice Cream Fork
424 A 17	Empire: Pie Fork
424 A 18	Empire: Salad Fork (Indv.)
424 A 19	Empire: Berry Fork
424 A 20	Empire: Nut Pick
424 A 21	Empire: Asparagus Tongs (Indv.)
424 A 22	Empire: Butter Spreader
424 A 23	Empire: Butter Spreader
424 A 24	Empire: Orange Knife
424 A 25	Empire: Breakfast Knife or Tea Knife
424 A 26	Empire: Fish Fork (Indv.)
424 A 27	Empire: Fish Knife (Indv.)
426 A 1	Empire: Berry Spoon
426 A 2	Empire: Preserve Spoon
426 A 3	Empire: Jelly Spoon
426 A 4	Empire: Sugar Spoon
426 A 5	Empire: Pickle Fork
426 A 6	Empire: Beef Fork
426 A 7	Empire: Butter Knife
426 A 8	Empire: Butter Knife
426 A 9	Empire: Butter Knife
426 A 10	Empire: Butter Knife
426 A 11	Empire: Butter Knife
426 A 12	Empire: Butter Knife
426 A 13	Empire: Sugar Sifter
426 A 14	Empire: Pie Server
426 A 15	Empire: Cheese Scoop
426 A 16	Empire: Cheese Scoop (Small)
426 A 17	Empire: Cheese Knife
426 A 18	Empire: Bon Bon Scoop
426 A 19	Empire: Sardine Server
426 A 20	Empire: Jelly Knife
426 A 21	Empire: Cracker Spoon
426 A 22	Empire: Cucumber Server
426 A 23	Empire: Caddy Scoop
428 A 1	Empire: Soup Ladle
428 A 2	Empire: Oyster Ladle
428 A 3	Empire: Gravy Ladle
428 A 4	Empire: Cream Ladle
428 A 5	Empire: Punch Ladle
428 A 6	Empire: Asparagus Server (Large)
428 A 7	Empire: Asparagus Server (Small)
428 A 8	Empire: Dish Gravy Spoon
428 A 9	Empire: Waffle Server
428 A 10	Empire: Ice Cream Server
428 A 11	Empire: Nut Spoon
428 A 12	Empire: Pea Spoon
428 A 13	Empire: Confection Spoon
430 A 1	Empire: Fish Knife (Large)
430 A 2	Empire: Fish Fork (Large)
430 A 3	Empire: Fish Knife (Small)
430 A 4	Empire: Fish Fork (Small)
430 A 5	Empire: Salad Spoon, L.H.
430 A 6	Empire: Salad Fork, L.H.
430 A 7	Empire: Salad Spoon, S.H.
430 A 8	Empire: Salad Fork, S.H.
430 A 9	Empire: Macaroni Server
430 A 10	Empire: Fried Oyster Server
430 A 11	Empire: Olive Fork, L.H.
430 A 12	Empire: Olive Spoon, L.H.
430 A 13	Empire: Lettuce Fork
430 A 14	Empire: Mustard Spoon
430 A 15	Empire: Salt Spoon
432 A 1	Empire: Cold Meat Fork
432 A 2	Empire: Cold Meat Fork (Large)
432 A 3	Empire: Toast Fork
432 A 4	Empire: Melon Knife/Fork
432 A 5	Empire: Cake Knife
432 A 6	Empire: Ice Cream Slicer
432 A 7	Empire: Crumb Scraper (Large)
432 A 8	Empire: Crumb Scraper (Small)
432 A 9	Empire: Nut Cracker
432 A 10	Empire: Sugar Tongs
432 A 11	Empire: Sugar Tongs (Small)
432 A 12	Empire: Bon Bon Tongs
432 A 13	Empire: Bon Bon Tongs (Small)
432 A 14	Empire: Ice Tongs
432 A 15	Empire: Sandwich Tongs
432 A 16	Empire: Butter Pick, 2-Tine
432 A 17	Empire: Ice or Pudding Spoon
433 A 1	Empire: Sardine Tongs
433 A 2	Empire: Olive Fork, S.H.
433 A 3	Empire: Saratoga Chip Server
433 A 4	Empire: Olive Spoon, S.H.
433 A 5	Empire: Cheese Fork
440 A 1	Louis XV: Meat Carving Knife, H.H.
440 A 2	Louis XV: Meat Carving Fork, H.H.

No.	Item
440 A 3	Louis XV: Game Carving Knife, H.H.
440 A 4	Louis XV: Game Carving Fork. H.H.
440 A 5	Louis XV: Steel, H.H.
440 A 6	Louis XV: Beefsteak Carving Knife, H.H.
440 A 7	Louis XV: Beefsteak Carving Fork, H.H.
440 A 8	Louis XV: Steel, H.H. (Small)
440 A 9	Louis XV: Table Knife
440 A 10	Louis XV: Medium Knife
440 A 11	Louis XV: Dessert Knife
442 A 1	Louis XV: Table Spoon
442 A 2	Louis XV: Table Fork
442 A 3	Louis XV: Dessert Spoon
442 A 4	Louis XV: Dessert Fork
442 A 5	Louis XV: Teaspoon
442 A 6	Louis XV: Teaspoon, Five O'clock
442 A 7	Louis XV: Soup Spoon
442 A 8	Louis XV: Coffee Spoon
442 A 9	Louis XV: Spoon
442 A 10	Louis XV: Boullion Spoon
442 A 11	Louis XV: Ice Cream Spoon
442 A 12	Louis XV: Egg Spoon
442 A 13	Louis XV: Orange Spoon
442 A 14	Louis XV: Salt Spoon (Indv.)
442 A 15	Louis XV: Oyster Fork
442 A 16	Louis XV: Ice Cream Fork
442 A 17	Louis XV: Berry Fork
442 A 18	Louis XV: Fork
442 A 19	Louis XV: Salad Fork (Indv.)
442 A 20	Louis XV: Nut Pick
442 A 21	Louis XV: Asparagus Tongs (Indv.)
442 A 22	Louis XV: Butter Spreader
442 A 23	Louis XV: Butter Spreader
442 A 24	Louis XV: Breakfast or Tea Knife
442 A 25	Louis XV: Breakfast or Tea Knife
442 A 26	Louis XV: Orange Knife
442 A 27	Louis XV: Knife
442 A 28	Louis XV: Fish Fork (Indv.)
442 A 29	Louis XV: Fish Knife (Indv.)
444 A 1	Louis XV: Berry Spoon
444 A 2	Louis XV: Preserve Spoon
444 A 3	Louis XV: Jelly Spoon
444 A 4	Louis XV: Sugar Spoon
444 A 5	Louis XV: Pickle Fork
444 A 6	Louis XV: Beef Fork
444 A 7	Louis XV: Sardine Server
444 A 8	Louis XV: Butter Knife
444 A 9	Louis XV: Butter Knife
444 A 10	Louis XV: Butter Knife
444 A 11	Louis XV: Butter Knife
444 A 12	Louis XV: Cheese Knife
444 A 13	Louis XV: Cheese Scoop (Large)
444 A 14	Louis XV: Cheese Scoop (Small)
444 A 15	Louis XV: Sugar Sifter
444 A 16	Louis XV: Pie Server
444 A 17	Louis XV: Jelly Knife
444 A 18	Louis XV: Bon Bon Scoop
444 A 19	Louis XV: Cracker Scoop
444 A 20	Louis XV: Caddy Scoop
444 A 21	Louis XV: Cucumber Server
446 A 1	Louis XV: Soup Ladle
446 A 2	Louis XV: Oyster Ladle
446 A 3	Louis XV: Gravy Ladle
446 A 4	Louis XV: Cream Ladle
446 A 5	Louis XV: Punch Ladle
446 A 6	Louis XV: Asparagus Server (Large)
446 A 7	Louis XV: Asparagus Server (Small)
446 A 8	Louis XV: Dish Gravy Spoon
446 A 9	Louis XV: Waffle Server
446 A 10	Louis XV: Ice Cream Server
446 A 11	Louis XV: Nut Spoon
446 A 12	Louis XV: Pea Spoon
446 A 13	Louis XV: Confection Spoon
448 A 1	Louis XV: Fish Knife (Large)
448 A 2	Louis XV: Fish Fork (Large)
448 A 3	Louis XV: Fish Knife (Small)
448 A 4	Louis XV: Fish Fork (Small)
448 A 5	Louis XV: Salad Spoon, L.H.
448 A 6	Louis XV: Salad Fork, L.H.
448 A 7	Louis XV: Salad Spoon, S.H.
448 A 8	Louis XV: Salad Fork, S.H.
448 A 9	Louis XV: Macaroni Server
448 A 10	Louis XV: Fried Oyster Server
448 A 11	Louis XV: Lettuce Fork
448 A 12	Louis XV: Olive Fork, L.H.
448 A 13	Louis XV: Olive Spoon, L.H.
448 A 14	Louis XV: Mustard Spoon
448 A 15	Louis XV: Salt Spoon
450 A 1	Louis XV: Cold Meat Fork (Large)
450 A 2	Louis XV: Cold Meat Fork
450 A 3	Louis XV: Toast Fork
450 A 4	Louis XV: Ice Cream Slicer (Large)
450 A 5	Louis XV: Ice Cream Slicer (Small)
450 A 6	Louis XV: Cake Knife
450 A 7	Louis XV: Melon Knife/Fork
450 A 8	Louis XV: Butter Pick, Twisted
450 A 9	Louis XV: Butter Pick, 2-Tine
450 A 10	Louis XV: Sugar Tongs
450 A 11	Louis XV: Sugar Tongs (Small)
450 A 12	Louis XV: Bon Bon Tongs
450 A 13	Louis XV: Bon Bon Tongs (Small)
450 A 15	Louis XV: Ice Tongs
450 A 16	Louis XV: Sandwich Tongs
450 A 17	Louis XV: Crumb Scraper (Large)
450 A 18	Louis XV: Crumb Scraper (Small)
450 A 19	Louis XV: Nut Cracker
451 A 1	Louis XV: Sardine Tongs
451 A 2	Louis XV: Olive Spoon, S.H.
451 A 3	Louis XV: Saratoga Chip Server
451 A 4	Louis XV: Fork
451 A 5	Louis XV: Fork
451 A 6	Louis XV: Olive Fork, S.H.
451 A 7	Louis XV: Cheese Fork
451 A 8	Louis XV: Terrapin Fork
451 A 9	Louis XV: Horse Radish Spoon
451 A 10	Louis XV: Fried Egg Server
458 A 1	Imperial Queen: Meat Carving Knife, H.H.
458 A 2	Imperial Queen: Meat Carving Fork, H.H.
458 A 3	Imperial Queen: Game Carving Knife, H.H.
458 A 4	Imperial Queen: Game Carving Fork, H.H.
458 A 5	Imperial Queen: Steel, H.H.
458 A 6	Imperial Queen: Beefsteak Carving Knife, H.H.
458 A 7	Imperial Queen: Beefsteak Carving Fork, H.H.
458 A 8	Imperial Queen: Steel, H.H. (Small)
458 A 9	Imperial Queen: Table Knife
458 A 10	Imperial Queen: Medium Knife
458 A 11	Imperial Queen: Dessert Knife
460 A 1	Imperial Queen: Table Spoon
460 A 2	Imperial Queen: Table Fork
460 A 3	Imperial Queen: Dessert Spoon
460 A 4	Imperial Queen: Dessert Fork
460 A 5	Imperial Queen: Teaspoon
460 A 6	Imperial Queen: Teaspoon, Five O'clock
460 A 7	Imperial Queen: Soup Spoon
460 A 8	Imperial Queen: Coffee Spoon
460 A 9	Imperial Queen: Spoon
460 A 10	Imperial Queen: Boullion Spoon
460 A 11	Imperial Queen: Ice Cream Spoon
460 A 12	Imperial Queen: Egg Spoon
460 A 13	Imperial Queen: Orange Spoon
460 A 14	Imperial Queen: Salt Spoon (Indv.)
460 A 15	Imperial Queen: Oyster Fork
460 A 16	Imperial Queen: Ice Cream Fork
460 A 17	Imperial Queen: Berry Fork
460 A 18	Imperial Queen: Fork
460 A 19	Imperial Queen: Salad Fork (Indv.)
460 A 20	Imperial Queen: Nut Pick
460 A 21	Imperial Queen: Asparagus Tongs (Indv.)
460 A 22	Imperial Queen: Butter Spreader
460 A 23	Imperial Queen: Butter Spreader
460 A 24	Imperial Queen: Breakfast or Tea Knife
460 A 25	Imperial Queen: Breakfast or Tea Knife
460 A 26	Imperial Queen: Orange Knife
460 A 27	Imperial Queen: Knife
460 A 28	Imperial Queen: Fish Fork (Indv.)
460 A 29	Imperial Queen: Fish Knife (Indv.)
462 A 1	Imperial Queen: Berry Spoon
462 A 2	Imperial Queen: Preserve Spoon
462 A 3	Imperial Queen: Jelly Spoon
462 A 4	Imperial Queen: Sugar Spoon
462 A 5	Imperial Queen: Pickle Fork
462 A 6	Imperial Queen: Beef Fork
462 A 7	Imperial Queen: Sardine Server
462 A 8	Imperial Queen: Butter Knife
462 A 9	Imperial Queen: Butter Knife
462 A 10	Imperial Queen: Butter Knife
462 A 11	Imperial Queen: Butter Knife
462 A 12	Imperial Queen: Cheese Knife
462 A 13	Imperial Queen: Cheese Scoop (Large
462 A 14	Imperial Queen: Cheese Scoop (Small
462 A 15	Imperial Queen: Sugar Sifter
462 A 16	Imperial Queen: Pie Server
462 A 17	Imperial Queen: Jelly Knife
462 A 18	Imperial Queen: Bon Bon Scoop
462 A 19	Imperial Queen: Cracker Scoop
462 A 20	Imperial Queen: Caddy Scoop
462 A 21	Imperial Queen: Confection Spoon
464 A 1	Imperial Queen: Soup Ladle
464 A 2	Imperial Queen: Oyster Ladle
464 A 3	Imperial Queen: Gravy Ladle
464 A 4	Imperial Queen: Cream Ladle
464 A 5	Imperial Queen: Punch Ladle
464 A 6	Imperial Queen: Asparagus Server (La
464 A 7	Imperial Queen: Asparagus Server (Sm
464 A 8	Imperial Queen: Dish Gravy Spoon
464 A 9	Imperial Queen: Waffle Server
464 A 10	Imperial Queen: Ice Cream Server
464 A 11	Imperial Queen: Nut Spoon
464 A 12	Imperial Queen: Pea Spoon
464 A 13	Imperial Queen: Cucumber Server
466 A 1	Imperial Queen: Fish Knife (Large)
466 A 2	Imperial Queen: Fish Fork (Large)
466 A 3	Imperial Queen: Fish Knife (Small)
466 A 4	Imperial Queen: Fish Fork (Small)
466 A 5	Imperial Queen: Salad Spoon, L.H.
466 A 6	Imperial Queen: Salad Fork, L.H.
466 A 7	Imperial Queen: Salad Spoon, S.H.
466 A 8	Imperial Queen: Salad Fork, S.H.
466 A 9	Imperial Queen: Macaroni Server
466 A 10	Imperial Queen: Fried Oyster Server
466 A 11	Imperial Queen: Lettuce Fork
466 A 12	Imperial Queen: Olive Fork, L.H.
466 A 13	Imperial Queen: Olive Spoon, L.H.
466 A 14	Imperial Queen: Mustard Spoon
466 A 15	Imperial Queen: Salt Spoon
468 A 1	Imperial Queen: Toast Fork
468 A 2	Imperial Queen: Cold Meat Fork
468 A 3	Imperial Queen: Cold Meat Fork (Larg
468 A 4	Imperial Queen: Melon Knife/Fork
468 A 5	Imperial Queen: Cake Knife
468 A 6	Imperial Queen: Ice Cream Slicer (Sm
468 A 7	Imperial Queen: Ice Cream Slicer (Lar
468 A 8	Imperial Queen: Butter Pick, Twisted
468 A 9	Imperial Queen: Butter Pick, 2-Tine
468 A 10	Imperial Queen: Sugar Tongs
468 A 11	Imperial Queen: Sugar Tongs (Small)
468 A 12	Imperial Queen: Bon Bon Tongs
468 A 13	Imperial Queen: Bon Bon Tongs (Sma
468 A 14	Imperial Queen: Nut Cracker
468 A 15	Imperial Queen: Ice Tongs
468 A 16	Imperial Queen: Sandwich Tongs
468 A 17	Imperial Queen: Crumb Scraper (Large
468 A 18	Imperial Queen: Crumb Scraper (Smal
469 A 1	Imperial Queen: Sardine Tongs

No.	Item
469 A 2	Imperial Queen: Fried Egg Server
469 A 3	Imperial Queen: Horse Radish Spoon
469 A 4	Imperial Queen: Fork
469 A 5	Imperial Queen: Saratoga Chip Server
469 A 6	Imperial Queen: Fork
469 A 7	Imperial Queen: Terrapin Fork
469 A 8	Imperial Queen: Cheese Fork
476 A 1	Heraldic: Table Spoon
476 A 2	Heraldic: Table Fork
476 A 3	Heraldic: Dessert Spoon
476 A 4	Heraldic: Dessert Fork
476 A 5	Heraldic: Teaspoon
476 A 6	Heraldic: Meat Carving Knife, H.H.
476 A 7	Heraldic: Meat Carving Fork, H.H.
476 A 8	Heraldic: Game Carving Knife, H.H.
476 A 9	Heraldic: Game Carving Fork, H.H.
476 A 10	Heraldic: Steel, H.H.
476 A 11	Heraldic: Table Knife
476 A 12	Heraldic: Dessert Knife
476 A 13	Heraldic: Coffee Spoon
476 A 14	Heraldic: Orange Spoon
476 A 15	Heraldic: Ice Cream Spoon
476 A 16	Heraldic: Egg Spoon
476 A 17	Heraldic: Boullion Spoon
476 A 18	Heraldic: Butter Spreader
476 A 19	Heraldic: Butter Spreader
476 A 20	Heraldic: Oyster Fork
476 A 21	Heraldic: Ice Cream Fork
476 A 22	Heraldic: Berry Fork
476 A 23	Heraldic: Fork
476 A 24	Heraldic: Nut Pick
478 A 1	Heraldic: Berry Spoon
478 A 2	Heraldic: Ice Cream Server
478 A 3	Heraldic: Cheese Scoop
478 A 4	Heraldic: Asparagus Server
478 A 5	Heraldic: Salad Spoon
478 A 6	Heraldic: Salad Fork
478 A 7	Heraldic: Soup Ladle
478 A 8	Heraldic: Olive Fork, L.H.
478 A 9	Heraldic: Olive Spoon, L.H.
478 A 10	Heraldic: Butter Knife
478 A 11	Heraldic: Butter Knife
478 A 12	Heraldic: Butter Knife
478 A 13	Heraldic: Pickle Fork
478 A 14	Heraldic: Sardine Server
478 A 15	Heraldic: Bon Bon Scoop
478 A 16	Heraldic: Cream Ladle
478 A 17	Heraldic: Caddy Scoop
478 A 18	Heraldic: Sugar Tongs
478 A 19	Heraldic: Sugar Tongs (Small)
496 A 1	Hyperion: Table Spoon
496 A 2	Hyperion: Table Fork
496 A 3	Hyperion: Dessert Spoon
496 A 4	Hyperion: Dessert Fork
496 A 5	Hyperion: Teaspoon
496 A 6	Hyperion: Meat Carving Knife, H.H.
496 A 7	Hyperion: Meat Carving Fork, H.H.
496 A 8	Hyperion: Game Carving Knife, H.H.
496 A 9	Hyperion: Game Carving Fork, H.H.
496 A 10	Hyperion: Steel, H.H.
496 A 11	Hyperion: Table Knife
496 A 12	Hyperion: Dessert Knife
496 A 13	Hyperion: Coffee Spoon
496 A 14	Hyperion: Egg Spoon
496 A 15	Hyperion: Orange Spoon
496 A 16	Hyperion: Oyster Fork
496 A 17	Hyperion: Berry Fork
496 A 18	Hyperion: Salad Fork or Ice Cream Fork
496 A 19	Hyperion: Pie Fork
496 A 20	Hyperion: Butter Spreader
496 A 21	Hyperion: Butter Spreader
496 A 22	Hyperion: Nut Pick
498 A 1	Hyperion: Berry Spoon
498 A 2	Hyperion: Ice Cream Server
498 A 3	Hyperion: Butter Knife
498 A 4	Hyperion: Butter Knife
498 A 5	Hyperion: Butter Knife
498 A 6	Hyperion: Salad Spoon
498 A 7	Hyperion: Salad Fork
498 A 8	Hyperion: Soup Ladle
498 A 9	Hyperion: Cheese Scoop
498 A 10	Hyperion: Sugar Spoon
498 A 11	Hyperion: Cream Ladle
498 A 12	Hyperion: Bon Bon Scoop
498 A 13	Hyperion: Pickle Fork
498 A 14	Hyperion: Sardine Server
498 A 15	Hyperion: Olive Fork, L.H.
498 A 16	Hyperion: Olive Spoon, L.H.
498 A 17	Hyperion: Sugar Tongs
498 A 18	Hyperion: Sugar Tongs (Small)
510 A 1	Bead: Teaspoon
510 A 2	Bead: Teaspoon, 5 O'clock
510 A 3	Bead: Dessert Spoon
510 A 4	Bead: Dessert Fork
510 A 5	Bead: Table Spoon
510 A 6	Bead: Table Fork
510 A 7	Bead: Soup Spoon
510 A 8	Bead: Coffee Spoon
510 A 9	Bead: Boullion Spoon
510 A 10	Bead: Ice Cream Spoon
510 A 11	Bead: Egg or Chocolate Spoon
510 A 12	Bead: Orange Spoon
510 A 13	Bead: Oyster Fork
510 A 14	Bead: Berry Fork
510 A 15	Bead: Fork
510 A 16	Bead: Salad Fork
510 A 17	Bead: Nut Pick
510 A 18	Bead: Butter Spreader
510 A 19	Bead: Chocolate or Egg Spoon
510 A 20	Bead: Terrapin Fork
510 A 21	Bead: Breakfast or Tea Knife
510 A 22	Bead: Breakfast or Tea Knife
510 A 23	Bead: Orange Knife
510 A 24	Bead: Knife
510 A 25	Bead: Fish Knife (Indv.)
510 A 26	Bead: Fish Fork (Indv.)
511 A 1	Bead: Berry Spoon
511 A 2	Bead: Preserve Spoon
511 A 3	Bead: Sugar Spoon
511 A 4	Bead: Jelly Spoon
511 A 5	Bead: Fork
511 A 6	Bead: Ice Cream Fork
511 A 7	Bead: Cold Meat Fork
511 A 8	Bead: Sardine Server
511 A 9	Bead: Butter Knife
511 A 10	Bead: Butter Knife
511 A 11	Bead: Sugar Sifter
511 A 12	Bead: Bon Bon Scoop
511 A 13	Bead: Caddy Scoop
511 A 14	Bead: Cheese Scoop
511 A 15	Bead: Cheese Knife
511 A 16	Bead: Pie Server
511 A 17	Bead: Jelly Knife
511 A 18	Bead: Waffle Server
511 A 19	Bead: Cracker Spoon
511 A 20	Bead: Cucumber Server
512 A 1	Bead: Soup Ladle
512 A 2	Bead: Oyster Ladle
512 A 3	Bead: Gravy Ladle
512 A 4	Bead: Cream Ladle
512 A 5	Bead: Macaroni Server
512 A 6	Bead: Asparagus Server
512 A 7	Bead: Dish Gravy Server
512 A 8	Bead: Nut Spoon
512 A 9	Bead: Fried Oyster Server
512 A 10	Bead: Ice Cream Server
512 A 11	Bead: Saratoga Chip Server
512 A 12	Bead: Confection Spoon
512 A 13	Bead: Mustard Spoon
512 A 14	Bead: Salt Spoon
512 A 15	Bead: Child's Spoon
512 A 16	Bead: Child's Fork
512 A 17	Bead: Child's Knife
512 A 18	Bead: Food Pusher
513 A 1	Bead: Fish Knife
513 A 2	Bead: Fish Fork
513 A 3	Bead: Salad Fork (Small)
513 A 4	Bead: Salad Spoon (Small)
513 A 5	Bead: Salad Fork (Large)
513 A 6	Bead: Salad Spoon (Large)
513 A 7	Bead: Cake Knife
513 A 8	Bead: Ice Cream Slicer
513 A 9	Bead: Olive Fork
513 A 10	Bead: Olive Spoon
513 A 11	Bead: Lettuce Fork
513 A 12	Bead: Cheese Fork
513 A 13	Bead: Sugar Tongs
513 A 15	Bead: Sugar Tongs (Small)
513 A 17	Bead: Ice Tongs
513 A 18	Bead: Sardine Tongs
513 A 19	Bead: Crumb Scraper
519 A 1	Light Oval Twist: Soup Ladle
519 A 2	Light Oval Twist: Gravy Ladle
519 A 3	Light Oval Twist: Cream Ladle
519 A 4	Light Oval Twist: Fish Knife
519 A 5	Light Oval Twist: Fish Fork
519 A 6	Light Oval Twist: Salad Spoon
519 A 7	Light Oval Twist: Salad Fork
519 A 8	Light Oval Twist: Asparagus Fork
519 A 9	Light Oval Twist: Nut Spoon
519 A 10	Light Oval Twist: Fried Oyster Server
519 A 11	Light Oval Twist: Waffle Server
519 A 12	Light Oval Twist: Dish Gravy Spoon
520 A 1	Light Oval Twist: Coffee Spoon
520 A 2	Light Oval Twist: Oyster Fork
520 A 3	Light Oval Twist: Orange Spoon
520 A 4	Light Oval Twist: Egg or Chocolate Spoon
520 A 5	Light Oval Twist: Fork
520 A 6	Light Oval Twist: Berry Fork
520 A 7	Light Oval Twist: Nut Pick
520 A 8	Light Oval Twist: Chocolate or Egg Spoon
520 A 9	Light Oval Twist: Butter Spreader
520 A 10	Light Oval Twist: Butter Spreader
520 A 11	Light Oval Twist: Teaspoon
520 A 12	Light Oval Twist: Berry Spoon
520 A 13	Light Oval Twist: Preserve Spoon
520 A 14	Light Oval Twist: Jelly Spoon
520 A 15	Light Oval Twist: Sugar Spoon
520 A 16	Light Oval Twist: Fork
520 A 17	Light Oval Twist: Sardine Server
520 A 18	Light Oval Twist: Bon Bon Scoop
520 A 19	Light Oval Twist: Sugar Sifter
520 A 20	Light Oval Twist: Butter Knife
520 A 21	Light Oval Twist: Butter Knife
520 A 22	Light Oval Twist: Pie Server
520 A 23	Light Oval Twist: Caddy Scoop
520 A 24	Light Oval Twist: Cheese Scoop
521 A 1	Light Oval Twist: Cold Meat Fork (Large)
521 A 2	Light Oval Twist: Cold Meat Fork
521 A 3	Light Oval Twist: Olive Fork
521 A 4	Light Oval Twist: Olive Spoon
521 A 5	Light Oval Twist: Lettuce Fork
521 A 6	Light Oval Twist: Salad Fork (Indv.)
521 A 7	Light Oval Twist: Sugar Tongs (Large)
521 A 8	Light Oval Twist: Sugar Tongs (Small)
521 A 9	Light Oval Twist: Bon Bon Tongs (Large)
521 A 10	Light Oval Twist: Bon Bon Tongs (Small)
521 A 11	Light Oval Twist: Sandwich Tongs
521 A 12	Light Oval Twist: Jelly Knife
521 A 13	Light Oval Twist: Ice Cream Server
521 A 14	Light Oval Twist: Ice Cream Slicer
521 A 15	Light Oval Twist: Mustard Spoon
521 A 16	Light Oval Twist: Salt Spoon
521 A 17	Light Oval Twist: Butter Pick, 2-Tine

521 A 18 Light Oval Twist: Butter Pick, Twisted
521 A 19 Light Oval Twist: Breakfast or Tea Knife
521 A 20 Light Oval Twist: Dessert Fork
521 A 21 Light Oval Twist: Dessert Spoon
521 A 22 Light Oval Twist: Food Pusher
548 A 1 Ivory: Berry Spoon
548 A 2 Ivory: Preserve Spoon
548 A 3 Ivory: Sugar Spoon
548 A 4 Ivory: Sugar Sifter
548 A 5 Ivory: Gravy Ladle
548 A 6 Ivory: Coffee Spoon
548 A 7 Ivory: Oyster Fork
548 A 8 Ivory: Ice Cream Fork
548 A 9 Ivory: Bon Bon Spoon
548 A 10 Ivory: Server
548 A 11 Ivory: Sardine Server
548 A 12 Ivory: Fried Egg Server
548 A 13 Ivory: Pie Server
548 A 14 Ivory: Nut or Ice Spoon
548 A 15 Ivory: Butter Pick
548 A 16 Ivory: Olive Fork
549 A 1 Ivory: Fish Knife
549 A 2 Ivory: Fish Fork
549 A 3 Ivory: Salad Spoon
549 A 4 Ivory: Salad Fork
549 A 5 Ivory: Asparagus Server
549 A 6 Ivory: Ice Cream Slicer
549 A 7 Ivory: Fried Oyster Server
549 A 8 Ivory: Lettuce Fork
601 A 1 Empire: Pie Server & Pie Forks
609 A 1 Louis XV: Ice Cream Slicer & Ice Cream Forks
658 A 1 Imperial Queen: Presentation Set In Presentation Box
658 A 2 Presentation Box
658 A 3 Imperial Queen: Presentation Set
763 A 1 Button Hook
763 A 2 Button Hook
763 A 3 Louis XV: Button Hook
763 A 4 Louis XV: Button Hook
763 A 5 Imperial Queen: Button Hook
763 A 6 Imperial Queen: Button Hook
763 A 7 Button Hook
763 A 8 Button Hook
763 A 9 Button Hook
763 A 10 Button Hook
763 A 11 Square Twist: Button Hook
763 A 12 Button Hook
763 A 13 Button Hook
763 A 14 Louis XV: Button Hook
763 A 15 Heraldic: Button Hook
763 A 16 Empire: Button Hook
763 A 17 Light Oval Twist: Button Hook
763 A 18 Button Hook
763 A 19 Imperial Queen: Button Hook
763 A 20 No. 8: Button Hook
763 A 21 No. 1: Button Hook
763 A 22 Button Hook
763 A 23 Button Hook
763 A 24 Square Twist: Button Hook
763 A 25 Hyperion: Button Hook
763 A 26 Lily Of The Valley: Button Hook
763 A 27 Button Hook
763 A 28 Glove Stretcher
763 A 29 Glove Stretcher
763 A 30 Glove Stretcher
763 A 31 Louis XV: Glove Stretcher
763 A 32 Imperial Queen: Glove Stretcher
775 A 117 Louis XV: Dresser Set
775 A 214 Louis XV: Dresser Set
775 A 215 Imperial Queen: Dresser Set
775 A 266 Imperial Queen: Dresser Set
775 A 319 Louis XV: Dresser Set
801 A 1 Jar
801 A 2 Jar
801 A 3 Jar
801 A 4 Jar
801 A 5 Jar
801 A 6 Jar
801 A 7 Jar
801 A 8 Jar
801 A 9 Jar
801 A 10 Jar
801 A 11 Jar
801 A 12 Jar
801 A 13 Jar
801 A 14 Jar
801 A 15 Jar
801 A 16 Paper Cutter
801 A 17 Paper Cutter
801 A 18 No. 1: Paper Cutter
801 A 19 Light Oval Twist: Paper Cutter
801 A 20 Heraldic: Paper Cutter
801 A 21 Lily Of The Valley: Paper Cutter
801 A 22 Louis XV: Paper Cutter
801 A 23 Imperial Queen: Paper Cutter
801 A 24 Empire: Paper Cutter
801 A 25 Book Mark
801 A 26 Book Mark
801 A 27 Book Mark
801 A 28 Paper Cutter
801 A 29 Paper Cutter
801 A 30 No. 1: Paper Cutter
801 A 31 Paper Cutter
801 A 32 Paper Cutter
801 A 33 Paper Cutter
801 A 34 Paper Cutter
801 A 35 Heraldic: Paper Cutter
801 A 36 Paper Cutter
801 A 37 Book Mark
801 A 38 Book Mark
801 A 39 Book Mark
801 A 40 Book Mark
801 A 41 Book Mark
801 A 42 Louis XV: Paper Cutter
801 A 43 Imperial Queen: Paper Cutter
801 A 44 Louis XV: Magnifying Glass
801 A 45 Louis XV: Magnifying Glass
801 A 46 Louis XV: Magnifying Glass
801 A 47 Louis XV: Magnifying Glass

Key to Abbreviations

H.H. = Hollow Handle
S.H. = Short Handle
L.H. = Long Handle
Indv. = Individual